SHROPSHIRE GEOLOGY

A Visitors' Guide

by

P. F. Phillips
and
J. R. Stratford

We wish to acknowledge:

for their inspiration and help over the years,
John Norton, Peter Toghill and Andrew Jenkinson;

John Norton, for permission to reproduce his excellent
fossil drawings (annotated 'JN');

the many co-operative landowners.

Photographs and Maps by P.F. Phillips
Except 'Wenlock Edge' - Cambridge University Collection of Air
Photographs: copyright reserved

Cover Photograph - The Stiperstones

Copyright 1995 P.F. Phillips and J.R. Stratford

Dedication: PFP: for my parents.

Published by:
Phillips Tutorials, Frogs Gutter, Minsterley, Shropshire. SY5 0NL.

Printed by
The Print Shop, 49 Church Street, Oswestry

Contents

How to use this guide .... page	5
A Brief Geological History of Shropshire	6
Geological Time Scale	7
Shrewsbury Area	13
Oswestry Area	15
The Old Lead/Barytes Mining Area	17
South Shropshire	23
The Onny Valley	31
Localities Map 36,	37
Church Stretton Area	39
Ludlow Area	46
The Wrekin and Ironbridge Gorge	51
Wenlock Edge	57
Glossary - Fossils	61
Glossary - Minerals	65
Glossary - Rocks	68

Frontispiece - The Stretton Hills from the Long Mynd

HOW TO USE THIS GUIDE

Shropshire is a wonderful county for geology. It was one of the most important areas for early students. For visitors interested in fossils, minerals and rocks, it offers some fascinating sites. Plus, of course, the lovely scenery of the southern and western hills.

This guide describes some of the more accessible and interesting areas for the visitor. It is intended primarily for those of you who have little knowledge of geology and to this end we have included a glossary, explaining some of the more technical terms used in the text and describing some of the fossils, minerals and rocks you may find and wish to collect.

Localities which may be visited have been grouped together into intineraries, each covering a small area and providing a convenient half or full day's excursion. However details are such that any locality can be visited separately. Alternatively you can select from the glossary a particular fossil, mineral or rock in which you are interested; suitable localities for finding the specimen are given.

Fossil finds may be identified by reference to British Palaeozoic Fossils, one of three volumes published by the Natural History Museum and a classic work of reference.

We strongly recommend the use of Ordnance Survey 1:50,000 maps of the area; sheet numbers are Landranger 126, 127, 137 and 138. Localities have been given standard six figure grid references, eg. (GR 385038).

Assuming that most of you will be travelling by car, we have given distances in miles (1 mile is approximately 1.6 km). For smaller measurements we have used yards, feet and inches; for those of you more familiar with the metric system, 1 yard is a little less than 1 metre, 1 foot equals 30 cm and 1 inch is 2.5 cm.

Most geological sites are on private land and mention of such sites in this guide does not give you the right to visit them. Landowners are, in our experience, very willing to allow visitors to enter their land, but do ask them first. Please observe the Conservation Code. Don't hammer wildly at the rock faces! Hunt in the scree and fallen rocks for your specimens; this requires much less effort and in most cases is far more rewarding. Open gates, don't climb on them, or on fences; and do not disturb farm animals. Remember that your behaviour will affect landowners' views of visitors who may follow.

We have made the directions and sketch maps as simple as possible, but if you have problems, please tell us so that we can improve the next edition. Likewise we include places to eat and drink; some will change, you may find others you'd recommend; and your comments will be welcome.

A BRIEF GEOLOGICAL HISTORY OF SHROPSHIRE

Just as Britain contains a greater variety of rocks within her boundaries than any other comparable area, so Shropshire, within Britain, is a county with perhaps a similarly wide range of rock types. Admittedly this range is mostly confined to the older end of the geological time scale; but so complete is the record here that the area has, ever since the early nineteenth century, become a classic one. Two geological periods - the Ordovician and the Silurian - take their names from ancient tribes once living in the area and were first studied in the county.

The Earth was formed about 4,600 million years ago and Shropshire has rocks dating back to over 600 million years. Throughout the brief account that follows, reference should be made to the geological time chart opposite.

The Pre-Cambrian
A major fault, or crack in the Earth's crust, known as the Church Stretton Fault, was active for an immense period of time, in terms of earthquakes and other forms of volcanic activity. Its line is marked by a range of hills, including The Wrekin (near Telford) and The Lawley and Caer Caradoc (near Church Stretton), which were formed in the Uriconian period some 600 plus million years ago. (See frontispiece photograph.) They consist of igneous rocks - lavas, ashes, etc.

To the west of the Church Stretton Fault, somewhat later, tens of millions of years of sedimentation took place, forming a series of rocks ranging from coarse conglomerates (shallow water or beach deposits) to fine-grained shales (deep water deposits). Fine-grained desert sandstones suggest a spell when the area was above sea level. These Longmyndian sediments form the Long Mynd itself and smaller outliers and are younger than the Uriconian. See photograph on page 8.

GEOLOGICAL TIME
(not to scale)

Era	Period	Years Ago (millions)	Shropshire Formations
Quaternary	Holocene / Pleistocene		Glacial Drift
		2 ---	
Tertiary or Cainozoic	Pliocene / Miocene / Oligocene / Eocene		Absent
		65 ---	
Mesozoic	Cretaceous		Absent
		135 ---	
	Jurassic		Shales
		190 ---	
	Triassic		New Red Sandstone
		225 ---	
	Permian		Absent
		280 ---	
Palaeozoic	Carboniferous		Coal Measures / Millstone Grit / Carboniferous Limestone
		345 ---	
	Devonian		Old Red Sandstone
		395 ---	
	Silurian		Aymestry Limestone / Wenlock Limestone / Wenlock Shale / Pentamerus Beds
		440 ---	
	Ordovician		Chatwall Sandstone / Hoar Edge Grit / Mytton Flags / Stiperstones Quartzite
		500 ---	
	Cambrian		Comley Limestone / Wrekin Quartzite
		570 ---	
	Pre-Cambrian		Longmyndian Sediments / Uriconian Volcanics

- Formation of the Earth ------- 4600 --------------------------------

The Long Mynd from the West

The Lower Palaeozoic
Around 570 million years ago the Cambrian period saw the first appearance of abundant life, as recorded by fossil remains. The trilobite, a many legged sea creature, is a good example. The period is not particularly well represented in Shropshire, producing mainly shales with thin layers of sandstone and limestones, indicating variation in the depth of the sea. Its onset is shown in the Wrekin area by the formation of a basal conglomerate.

In the Ordovician which followed, shellfish (brachiopods) and other sea creatures abounded and their remains are plentiful in the shales and sandstones of this period. Other major formations occur in the Stiperstones area, such as a quartzite - a hard slightly metamorphosed sandstone - and a dark fine-grained sandstone, known as the Mytton Flags, which is important because of the later injection of mineral veins, which gave rise to the lead mining in Shropshire. A resurgence of volcanic activity occurred, particularly over the border in Snowdonia. During the subsequent Silurian (440 to 395 million years ago) Shropshire saw a classic ebb and flow of the sea, giving rise to a wonderful series of rocks which is world renowned. This array is complicated by the Church Stretton Fault, for the depth of the sea was very different on either side of it.

Some of the older hills remained as islands; and conglomerates mark the Silurian shoreline, eg. the Kenley Grit in the east and a basal conglomerate all around the southern end of the Long Mynd. Waters close to the shoreline provided a home for numerous brachiopods and trilobites; whilst graptolites are found in the more fine-grained shales, laid down in the deeper sea.

Warm shallow waters provided limestones, eg the fossiliferous limestone of Wenlock Edge (Wenlock Limestone) and the rather harder limestone of View Edge (Aymestry Limestone) which is packed with the large *Kirkidium knighti* and the smaller *Atrypa reticularis* brachiopods and stem sections of crinoids (sea lilies). The hard rocks form the striking scarp ridges of South Shropshire, while the intervening shales form beautiful vales, eg Hope Vale. (See photograph on page 11.)

Most of the Siliurian rocks are a happy hunting ground for the fossil collector and the rocks themselves are both attractive and varied. They are, perhaps, the assemblage which has made the county one of the geologists' points of pilgrimage.

The Upper Palaeozoic
It seems likely that the close of the Silurian and the mountain building earth movements which marked it, saw most of Shropshire above sea level. The climate was tropical and the Devonian rocks are confined to the east and far

CROSS-SECTION IN SOUTH SHROPSHIRE

Wenlock Edge

south, which was probably an area of fresh water lakes and river estuaries. A thin stratum at the base of Shropshire's Devonian is one of the world's most famous; it is the Ludlow Bone Bed and consists of small pieces of the scales and spines of primitive fish in a soft sandy matrix. So overworked has it been that, in the cliffs at Ludlow, it is recognisable only as a deep notch cut out of the face by collectors. For the rest of the Devonian, large areas are covered with "Old Red Sandstone", a red-brown or mottled sandstone, with layers of other sandstones and limestones. It forms the high plateau of the Clee Hills and Clun Forest.
This was followed by the Carboniferous, a period of equatorial forests and deltas,

with limited but economically important representation in Shropshire.
Mountain Limestone is found to a small extent near the Wrekin, and on a much larger scale at Llanymynech on the Welsh border. Coal was deposited and formerly mined, on the Clee Hills and in the Coalbrookdale area, where it provided fuel for Abraham Darby's furnaces that are recognised as the crucible of the Industrial Revolution; and also just south of Shrewsbury in the Hanwood and Pontesford areas. Millstone grit, an important Carboniferous formation in the north of England, is exposed in the Oswestry district, which again has a history of coal mining.

The Mesozoic

The Permian and Triassic Periods (280 to 190 million years ago) mark the transition from the Palaeozoic to the Mesozoic. Most of Britain was at this time hot desert and red desert sandstones formed, ie the "New Red Sandstone", composed of wind-rounded grains, readily discernible with a hand lens. In cuttings and cliff faces, eg at Bridgnorth, the layers of rock display dune bedding, swept up and down in curves by the desert winds. Mixed in with the sandstones are marls. Most of Shropshire north and east of the Severn is of New Red Sandstone. The generally flat plain is broken by a range of low sandstone hills (Nesscliffe, Harmer Hill, etc.). A small outcrop of Jurassic rocks, sandstones and clays, occurs in the north of the county around Prees, but the Cretaceous is absent; although it is possible that the Chalk sea did once extend over the whole of England, the rocks having been worn away.

The Tertiary

The last 65 million years of geological history is mostly unrepresented here in Shropshire. The last major geological event was the series of ice ages of the last 2 million years. Ice sheets and glaciers spread great depths of boulder clay over the Shropshire plain and many of the valleys, bringing rocks from all over the north west of Britain and North Wales. Also massive diversion of rivers took place, including that of the Severn, which before glaciation was a tributary of the Dee.

ITINERARIES

The localities described below are shown on the map on pages 36 and 37. They are grouped together into itineraries according to area, but there is some degree of overlap and certain localities may be mentioned more than once.
Ordnance Survey maps referred to are the 1:50,000 Landranger series.
Suggested intineraries are:

A: Shrewsbury Area. OS sheet 126.
B: Oswestry Area. OS sheet 126.
C: The Old Lead/Barytes Mining Area. OS sheets 126 & 137.
D: South Shropshire. OS sheet 137.
E: The Onny Valley. OS sheet 137.
F: Church Stretton Area. OS sheet 137.
G: Ludlow Area. OS sheet 137.
H: The Wrekin and Ironbridge Gorge. OS sheet 127.
J: Wenlock Edge. OS sheets 127, 137 (and 138).

Itinerary A: SHREWSBURY AREA

Shrewsbury itself sits near the edge of the great New Red Sandstone plain that stretches through North Shropshire, Cheshire and Lancashire. Close by, in an arc to the south of the town, were coal mines, now all closed (villages like Hanwood and Annscroft). To the east is the Coalbrookdale industrial area (Telford) and The Wrekin.

The immediate vicinity of Shrewsbury is not the most exciting area for geology in Shropshire, but it is a convenient centre and most itineraries are within easy reach. When in the town do visit Rowley's House Museum in Barker Street; where there are some excellent geological exhibits and a well stocked souvenir shop.

A1. Callow Quarry (GR 385048)
Ordovician Flagstones and Dolerite.
From Shrewsbury follow the A488 to Minsterley. Turn left on the road signposted Habberley, after ½ mile, left again into the quarry entrance. This is a working quarry, so take care where you park and ASK PERMISSION to enter.

The predominant rock is a dark grey flagstone, known as Mytton Flags - the same rock in which you find the old lead mines at Snailbeach and The Bog. In fact old lead workings have been uncovered here. Walk through to the northern end where there has been recent quarrying activity. Note the steeply dipping bedding planes of the Mytton Flags; they are often contorted. An intrusion of dolerite has cut through the quarry and this has caused baking and hardening of the flags, producing a rock suitable for roadstone. Samples of this dolerite are plentiful, especially around the prominent mound in the centre of this area of the quarry.

From Minsterley you can head south to Snailbeach and The Bog - see Itinerary C: The Old Lead/Barytes Mining Area; or alternatively you can return northwards towards Shrewsbury; turning right on the new A5 towards Telford and The Wrekin - see Itinerary H: The Wrekin and Ironbridge Gorge; or turning left on the A5 towards Oswestry to visit locality A2 and then onto Itinerary B: Oswestry Area.

A2. Nesscliffe (GR 384193)
Triassic New Red Sandstone.
From Shrewsbury take the A5 to the west. The village of Nesscliffe is about 8 miles from Shrewsbury. Past the petrol station on the left is The Old Three Pigeons Inn; opposite here take a minor road signposted to Hopton and Valeswood. Immediately on your right you will see a wooden gate, with a notice Nesscliffe Hill Country Park; this is the way upto the red sandstone cliffs. Park in the small layby almost opposite the gate, or ½ mile further on down the road. The rock of Nesscliffe Hill is New Red Sandstone, made up of wind-rounded grains of quartz, stained red by iron oxides. It was laid down under desert conditions, in Triassic times, some 200 million years ago. There are no fossils as life would not have been plentiful and anyway the desert conditions were not suitable for preservation. Follow the signs to Kynaston's Cave (there's a map and information just inside the gate). Examine the vertical cliff face opposite the second flight of wooden steps up to the cave, you will notice the varied bedding planes, known as cross bedding, produced by the deposition of the sand by the desert winds. Retrace your steps down from the cave to the main path; 80 yards further on is a wooden gate on the left and beyond this gate is an even more inpressive cliff face. Continue on the circular route to visit other interesting sites in the park or return directly to your car.

Continue westwards on the A5 for Itinerary B: Oswestry Area, or return towards Shrewsbury for any of the other itineraries.

Itinerary B: OSWESTRY AREA

B1. Nesscliffe
See Itinerary A: Shrewsbury Area, locality A2.
Continue north west on the A5 towards Oswestry and head for the town centre. There is an excellent market here on Wednesdays.

B2. The "Old Racecourse" (GR 255298)
Carboniferous Millstone Grit, sandstone and conglomerate and brachiopods.
This is not an easy locality and is for those with some time and patience. From Oswestry take the B4580 to Rhydycroesau; after a little more than 2 miles, at the crossroads at the top of the long hill, turn left (this is the site of the old racecourse). Continue southwards for ¾ mile until you see a narrow metalled road forking acutely to the right; take this road and after about ¼ mile, at a hairpin bend to the left, you'll see a cattle grid and house ahead and a track to the right. Park unobstructively here. Walk along the track and you will see the old quarry, partly obscured by trees on the right. There is a steep face and you will have to scramble about halfway up this to find a band of brachiopods (*Spirifer, and Productus*), in brown leached sandstone.

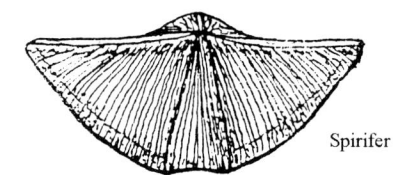

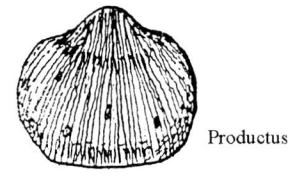

Spirifer   Productus

The rocks dip from top left to bottom right and are of Millstone Grit (Carboniferous) age. There are distinctive bands of attractive conglomerates containing white quartz pebbles.

B3. Llanymynech Quarry (GR 265217)
Carboniferous mountain limestone, brachiopods, corals and crinoids.
The huge cliff face at Llanymynech is visible from miles around. The quarry is no longer working and large parts of it form a nature reserve containing distinctive lime-loving plants.
Drive south from Oswestry on the B5069 and then the A483. After about 4 miles you will reach the village of Pant; note the Cross Guns pub on the right (west). About 200 yards further south is a small crossroads (GR 273221). Take the lane to the right (west). Drive to the end of this lane and park without obstructing any gate. You enter the quarry on a signed footpath. As you walk along note the

massive high rock face with very apparent, almost horizontal, bedding planes and vertical joints. This gives the massive appearance typical of mountain limestone.

Walking along the path you will find on the left an old engine house. Behind it is an artificial incline, up and down which an endless line of quarry waggons would have travelled, taking the rock down from the quarry. Observe the sleepers. You will meet a stile; cross it. 200 yards further on, take the path to the right (post with an acorn on it) which leads you into the quarry. Entering the quarry you will see below you a vast excavated bowl which was the main exit. There are two good fossiliferous areas.

(a) Ahead, climb **WITH CARE** to about 8 feet from the top of the cliff face (there is only one obvious way up). You will find a band up to 2 feet thick containing corals (*Lithostrotion sp*). There are also clusters of attractive calcite crystals ("nailhead" and "dogtooth"). These items, along with crinoid ossicles and brachiopods, may also be collected from the scree at the foot of the cliff.

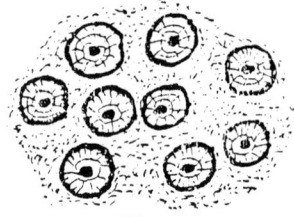

Lithostrotion

(b) In the coombe to the left - enter via style and rusty old fence - you will find plenty of complete brachiopods.

Return to the A483 and continue south for 2½ miles to the village of Four Crosses; fork left onto the A4393. After a further 2½ miles you cross the river Severn (narrow bridge with traffic lights). ½ mile after the bridge turn right and then left to Criggon, a small village nestling at the foot of Breidden Hill and covered in rock dust from the working quarry

B4. **Breidden Hill** (GR 292143)
Dolerite.

The public road passes through the lower part of the quarry works. The rock is a very tough, attractive green olivine dolerite, used for roadstone and also as a decorative finish for paths and driveways. Some samples of the rock contain small white specks, which are "rotting" crystals of feldspar.

Visit:
Ellesmere - Shropshire's Lakeland. The Meres Visitor Centre, new and purpose built, has displays which interpret the natural history and geology of the area. Souvenirs and book shop.

Itinerary C: THE OLD LEAD/BARYTES MINING AREA

This itinerary begins at the village of Snailbeach, 10 miles southwest of Shrewsbury. Take the A488 through Pontesbury and Minsterley and about 1¼ miles past Minsterley turn left to Snailbeach.

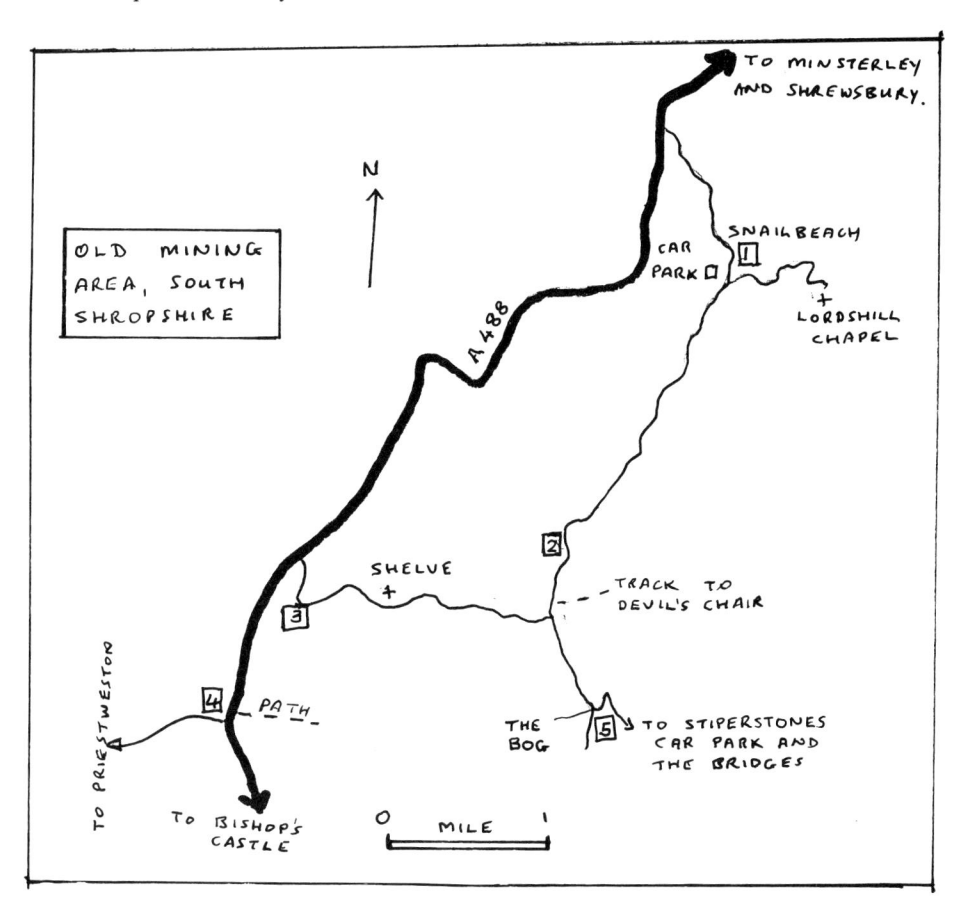

C1. Snailbeach (GR 375022)
Here was one of the most important lead mines in England; peak activity being in the second half of the nineteenth century. It was worked with Cornish engines of an advanced technology and had reached a depth of 1650 feet by 1911 when deep mining finished and the workings were allowed to flood. There was sporadic

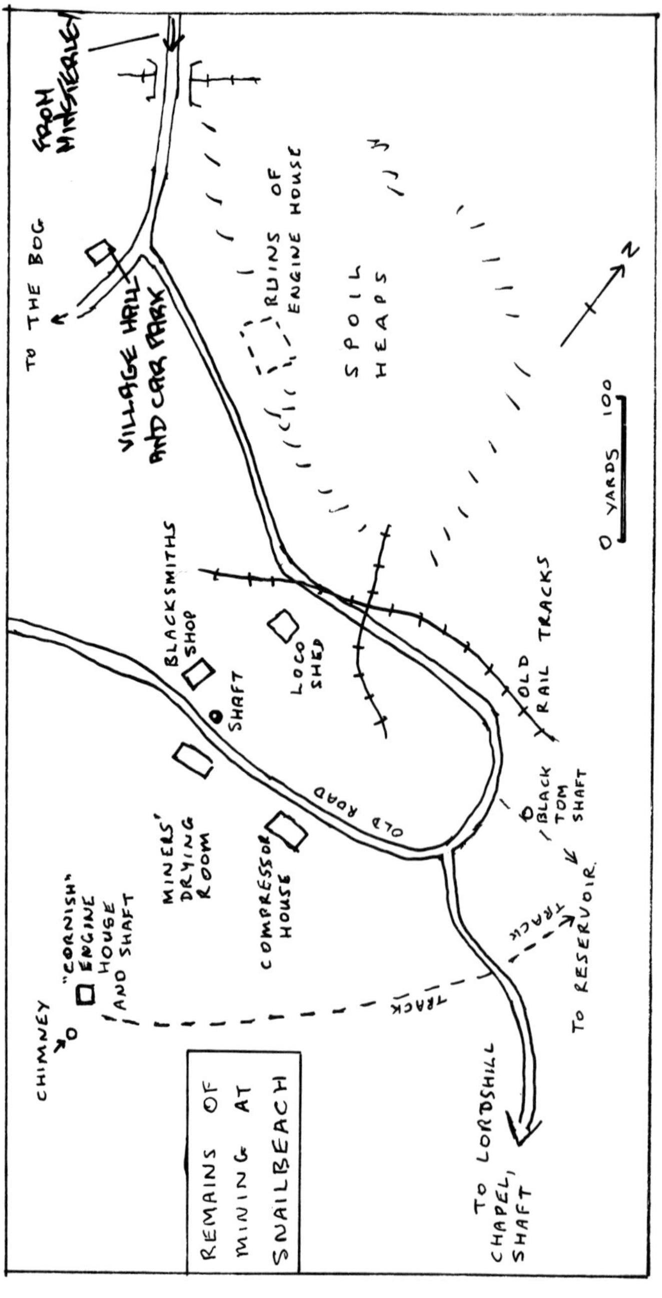
REMAINS OF MINING AT SNAILBEACH
FROM MINSTERLEY
TO THE BOG
VILLAGE HALL AND CAR PARK
RUINS OF ENGINE HOUSE
SPOIL HEAPS
N
0 YARDS 100
BLACKSMITHS SHOP
SHAFT
LOCO SHED
OLD RAIL TRACKS
OLD ROAD
BLACK TOM SHAFT
TO RESERVOIR
TRACK
TRACK
MINERS' DRYING ROOM
COMPRESSOR HOUSE
"CORNISH" ENGINE HOUSE AND SHAFT
CHIMNEY
TO LORDSHILL
TO CHAPEL, SHAFT

extraction of barytes from veins at or near the surface until 1955 when even this finally ceased. Today, the settlement pattern, plus chimneys, engine houses and spoil heaps witness the industrial past. At the time of writing the area is undergoing large scale making-safe and conservation work.

Park near the village hall (GR 373023). The large white spoil heaps, visible from many miles, are of calcite, barite and quartz, with small amounts of galena and sphalerite. Do not trespass onto the tips. Observe the old railway engine sheds on the right of the road. Further back (see plan) on the right are other buildings, including a compressor house and blacksmith's shop and nearby is the collapsed winding gear beside the derelict shaft (George's Shaft), which has recently been capped. Halfway up the wooded hill behind you will see a large engine house and chimney, which can be examined by taking the Lordshill road and then the track to the right - see map.. The engine house has recently undergone considerable restoration and the adjacent shaft has been made safe by a metal grill. This shaft is almost 1400 feet deep and a horizontal adit, the entrance to which can be seen between the compressor house and the miners' drying room, enters it some 350 feet below the surface. A little way up the valley, past the site of Black Tom Shaft, is the large reservoir, completed in 1872, which served the mine, providing water for the steam engines and for ore separation. Further along the track on the right you will see two adits, the second one of which is Perkins Level, which was used for the extraction of barytes in the twentieth century.

C2. Tankerville Mine (GR 355996)
Drive south from Snailbeach for about 2½ miles, passing through the village of Stiperstones, until you see the pottery at Tankerville. Park carefully here - ask permission at the pottery and do not obstruct the track. Walk 100 yards further south along the road to a footpath and stile on the right. A little way along the path here you will have a clear view of the very fine derelict engine house and chimney. See colour photograph on page 21. Do not trespass any closer.
Serious mining took place here between 1870 and 1884, working the richest vein in Shropshire, by means of the deepest shaft in the ore field.

C3. Ladywell Mine (GR 327992)
From Tankerville, continue south along the road to Pennerley and turn right (west), signed Shelve. Drive through Shelve and after ½ mile, on your left, you will see the remains of Ladywell engine house, a dramatic structure mimicing a Norman keep and a prominent local landmark. Ladywell is an example of a commercial mining failure; a great deal of money was invested here during its

short working life - 1871 to 1882 - but only 835 tons of ore were raised during these 11 years.

C4. The Grit Mines (GR 326981)
Galena, sphalerite, barite, calcite and quartz.
Continue to the A488 and turn left; in about 1 mile you will see The More Arms on your left and a little further on, on the right (west side of the road), just before the turning to Priest Weston, the White Grit engine house, precariously derelict. Opposite is a stile and footpath leading up the hill to the sites of East Grit and Old Grit mines. There is an interesting assemblage of shafts (be careful!) and waste tips, with the minerals quartz, calcite, barite, galena and sphalerite. This is the part of the orefield where mining has its longest history. There is evidence that lead ore was extracted by the Romans and probably also in the Middle Ages. There are records that mining was taking place here as early as 1678, reaching its peak around 1840. This once active mine has lain neglected since 1870.

Drive back the way you've come, to Pennerley crossroads, and turn right. After a mile you can park at The Bog.

C5. The Bog Mine (GR 356979)
Barite, calcite and quartz.
Once one of the greatest lead and barite mines in the county, this is now a desolate area. The remaining buildings were demolished over the last 20 years and extensive reclamation has disguised the scars of industrialisation. But there are enough remains such as building foundations, reservoirs, and the site of the old tramway to engage one's interest. In the repaired but roofless Miners' Institute are maps showing features of interest. Lead occurred as galena; and barite and sphalerite were also mined. You will also find calcite and quartz. Small amounts of metals such as cadmium and arsenic have produced a particularly toxic soil which poses problems for re-vegetating. Note the trial patches where researchers are assessing different varieties of grasses. Lead mining ceased here in 1883, but the site was worked for barytes during and after the first world war, at which time there was an aerial ropeway to carry the ore to a crushing mill at Minsterley.

Tankerville chimney, Locality C2;
Snailbeach waste tips in background, Locality C1

Hillend, Locality D2;
roadside site for collecting *Pentamerus* ("Government Rock")

Billingham ventile®

all-weather clothing *plus*

It's been keeping people warm, dry and alive for 50 years

Visit the Billingham Ventile Centre to find out more. 100% cotton Ventile, stormproof and breathable.

The Billingham Ventile Centre
9 High Street, Church Stretton,
SY6 6BU   Tel: 0694 724491

®Ventile is the registered Trade Mark of Courtaulds plc

**Billingham Ventile
Clothing Centre
Church Stretton**

*Open Tue - Sat from 9.30am.
Stocks a full range of Ventile Jackets, Gaiters and Overtrousers along with fleece jumpers, Boots and a host of outdoor accessories.
Telephone: 01694 724491*

Discover geology with your museum service

On Wenlock Edge
From tropical sea
to limestone escarpment
Much Wenlock Museum
Open April to October
Telephone: 01952 727773

Reading the Rocks
Celebrating Ludlow's contribution
to international geology
Ludlow Museum
Open April to October
Telephone: 01584 875384

The Meres & Mosses
A landscape shaped by ice
The Meres Visitor Centre
Open April to October
Telephone: 01691 622981

For more information contact the
County Curator for Natural Sciences
at Ludlow Museum Offices,
47 Old Street, Ludlow, SY8 1NW.
Telephone: 01584 873857

SHROPSHIRE
COUNTY COUNCIL
Leisure Services Department

Itinerary D: SOUTH SHROPSHIRE

This excursion covers the Stiperstones, the southern end of the Longmynd and two disused fossiliferous quarries. It may be taken as an extension of the previous Itinerary C, if time permits, or as a separate entity.

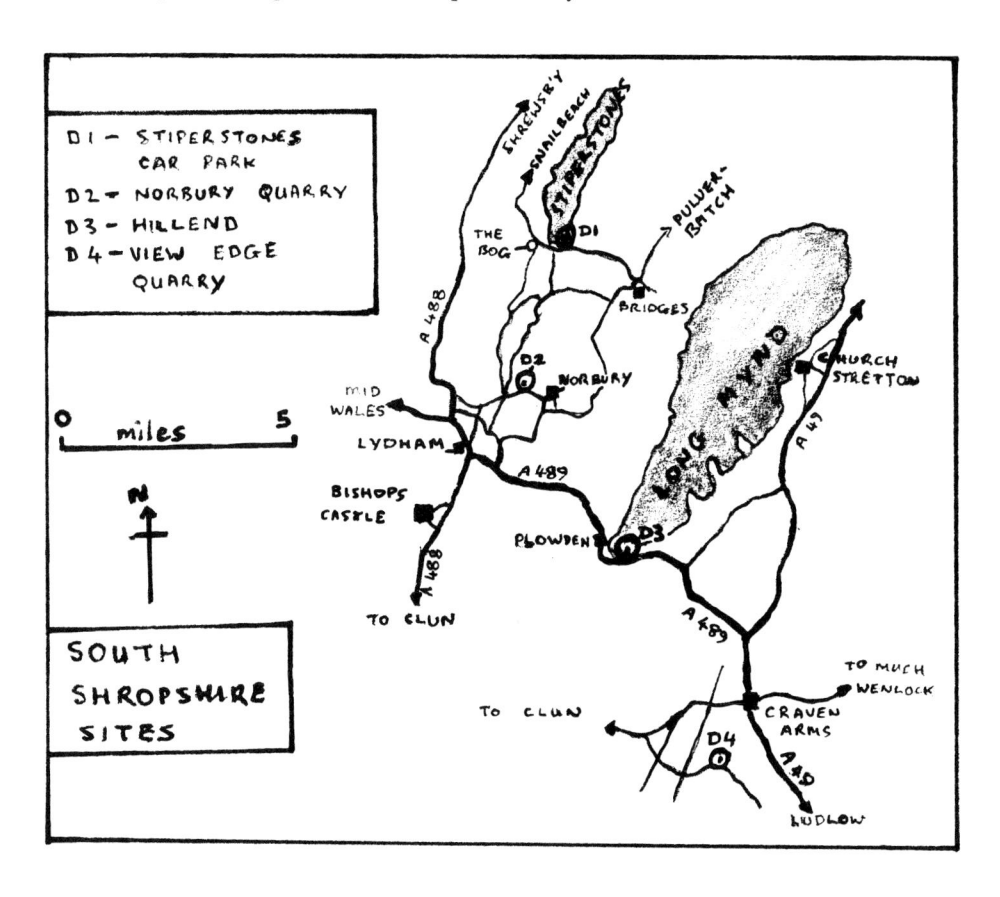

From Shrewsbury take the A488 south, through Pontesbury and Minsterley and almost 5 miles past Minsterley turn left at the crossroads (GR 326996); after 2¼ miles (having passed Shelve church on your left) you will reach the T-junction at Pennerley.

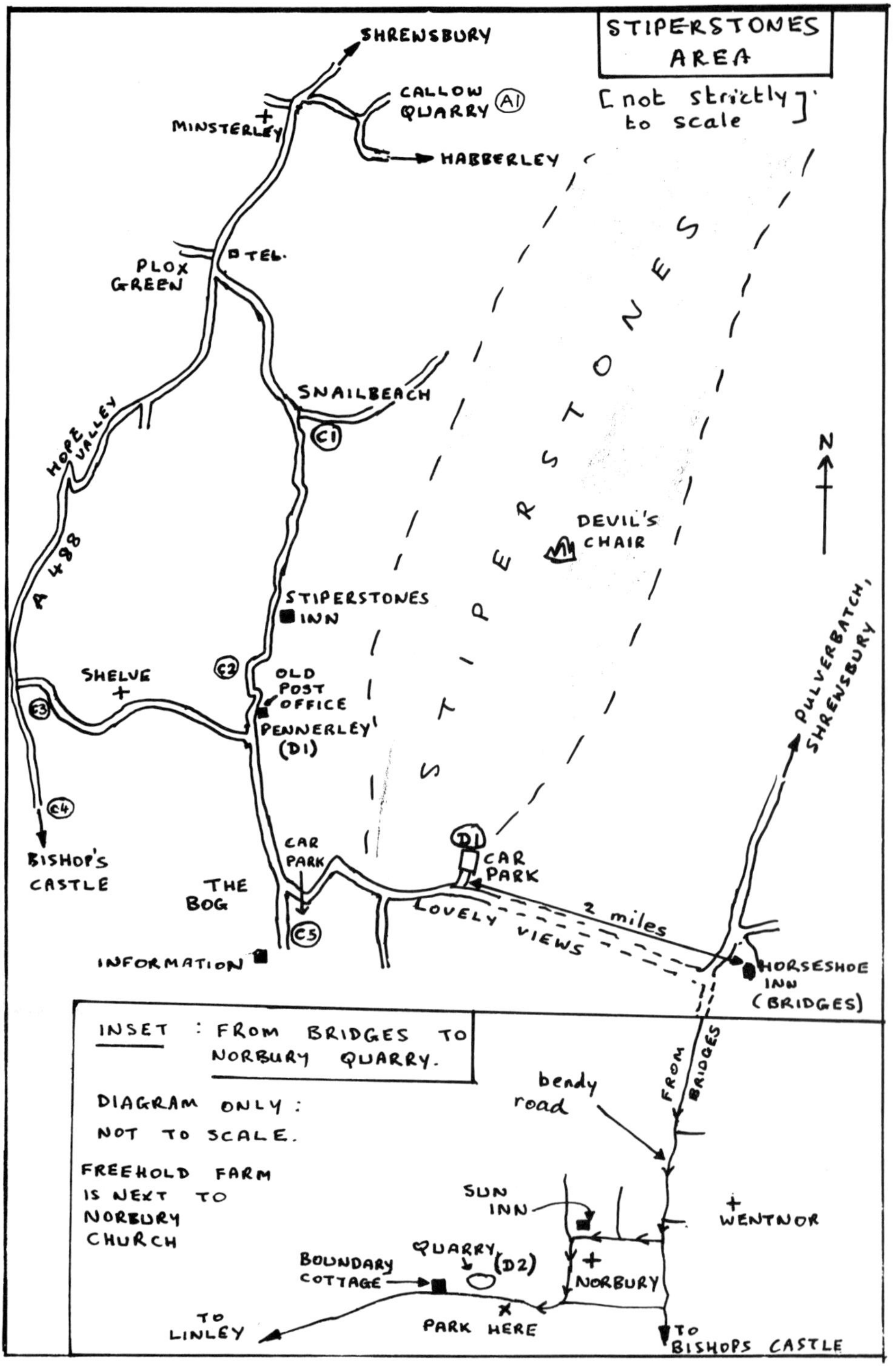
STIPERSTONES AREA
[not strictly to scale]
SHREWSBURY
CALLOW QUARRY
A1
MINSTERLEY
HABBERLEY
TEL.
PLOX GREEN
STIPERSTONES
SNAILBEACH
C1
HOPE VALLEY
A 488
N
DEVIL'S CHAIR
STIPERSTONES INN
C2
OLD POST OFFICE
SHELVE
C3
PENNERLEY
(D1)
PULVERBATCH, SHREWSBURY
C4
BISHOP'S CASTLE
D1
CAR PARK
CAR PARK
THE BOG
2 miles
LOVELY VIEWS
C5
INFORMATION
HORSESHOE INN (BRIDGES)
INSET : FROM BRIDGES TO NORBURY QUARRY.
DIAGRAM ONLY : NOT TO SCALE.
FREEHOLD FARM IS NEXT TO NORBURY CHURCH
bendy road
FROM BRIDGES
SUN INN
WENTNOR
QUARRY
(D2)
NORBURY
BOUNDARY COTTAGE
TO LINLEY
PARK HERE
TO BISHOPS CASTLE

D1. The Stiperstones (GR 368992/366981)
Ordovician quartzite and conglomerate.
A craggy ridge of quartzite (a toughened sandstone) some 500 million years old. The quartzite is a pale creamy, almost white colour and there are occasional bands of conglomerate with small rounded quartz pebbles, indicating ancient beach deposits. The rock layers are tilted at an angle of 70 degrees - you can see the lines of dip when you climb up. On top are the distinctive outcrops of rock, or tors, the most northerly one being named The Devil's Chair and the most southerly Cranberry Rock. There are two starting points from which you can reach the ridge.

From Pennerley (GR 353988) you can take a rough track up by Pennerly Old Post Office (200 yds north of the T- junction); drive up and you will see signs to the Devil's Chair. Parking is limited at the end of the track. From here you can climb up the flanks of the Stiperstones, noting the remains of miners' cottage walls and gardens. From the Devil's Chair, on a clear day, you can see Snowdonia, Cader Idris, Plylimmon, the Malverns and the Wrekin.

Alternatively turn south from Pennerly to The Bog (GR 356979); here, turn left (signed Bridges) and, at the Y junction on the top of the hill, left again. There is a large car park (GR 370977) by the first cattle grid. From here you can stroll up to Cranberry Rock and along the crest of the Stiperstones to the Devil's Chair.

D2. Norbury Quarry (GR357928)
Silurian Pentamerus beds, brachiopods, corals and crinoids.
The village of Norbury is 4 miles due south of the Stiperstones, west off the Bridges to Bishops Castle road. Stop at Freehold Farm, next to the church (GR 364928) and ASK PERMISSION from Mr and Mrs Edwards. The quarry is about 600 yds west along the lane towards Linley, on the north side. Park off the road - only room for 2 or 3 cars - and enter the quarry via a gate. Turn right inside the gate. Please CLOSE THE GATE.

This little known, small, disused and somewhat overgrown quarry is rich in fossils. The rock, known as the Pentamerus Beds, occurs towards the base of the Silurian, (ie. around 430 million years old) and is technically a sandstone, although the quantity of shell remains in places have turned some of it into virtually a limestone. However in this quarry the fossilised shells have frequently been dissolved away by percolating water and the fossils appear as hollows in the sandstone.

The main fossils here are: (i) *Pentamerus oblongus*; a brachiopod. Whole shells are not readily found, what is more usual is a cross-section, showing the rim of

Pentamerus Oblongus - "Government Rock"

the shell and the median septum (a dividing wall in the centre of the shell). which together can frequently be seen as an arrow shape, similar to the government mark and the convicts' broad arrow, hence the term "Government Rock"
(ii) *Streptelasma sp*; a very pretty small solitary, cone-shaped, coral, with longitudinal ridges.
(iii) Sections of crinoid stem or single ossicles, with the appearance in cross-section of a washer. In Ordovician-Silurian times this area was a shallow sea around the area of the Long Mynd. In this quarry you may also find samples of conglomerate, a concrete-like mixture of green, red and other pebbles, marking the shoreline and deposited just prior to the Pentamerus Beds. The pebbles

Streptelasma

were washed into the sea from the Long Mynd. Samples of purple Longmyndian sandstone may also be seen.

D3. Hillend (GR 396876)
Silurian Pentamerus beds, brachiopods, trilobites, Silurian conglomerate and Longmyndian Sandstone.
This is a roadside exposure on the north side of the A489 between Lydham and Craven Arms, just east of Plowden and at the southern end of the Long Mynd. You can park on the wide grass verge on the opposite (south) side, but do take care as traffic travels fast here. DO NOT HAMMER the bank; look for fossils in the scree below. You will see many examples of the brachiopod *Pentamerus oblongus* (see D2. above), indicating that once again these are the Pentamerus Beds. You will also find other brachiopods and, if you're lucky, a trilobite. See colour photograph on page 21. On the hillside above this exposure you will find outcrops of Silurian basal conglomerate and, a little higher up, Pre-Cambrian Longmyndian Sandstone.

D4. View Edge Quarry, near Craven Arms (GR 426806)
Aymestry Limestone, brachiopods and crinoids.
This location is worth visiting for its view as well as its fossils. It is on top of a section of Wenlock Edge, separated from the main ridge by the river Corve. From Craven Arms take the B4368(4367) towards Clun; after ¾ mile turn left onto a minor road, signposted Broome, Leintwardine (B4367). Two miles further on at the crossroads turn left again and drive up the narrow hill road. Stop at the farm on the top of the hill on the left and ASK PERMISSION. Entrance to the quarry is ¼ mile further on, on the left. Park tidily by the gate. Walk into the quarry (northwards) and you will find several rock faces of Aymestry Limestone. Search the scree for specimens; you have to collect from the scree as the rock is too hard to hammer. It is packed with fossils, particularly brachiopods (eg *Kirkidium* and *Atrypa*) and crinoid ossicles (ie sections of the stems of "sea lilies"), both circular and 5-point star shaped.

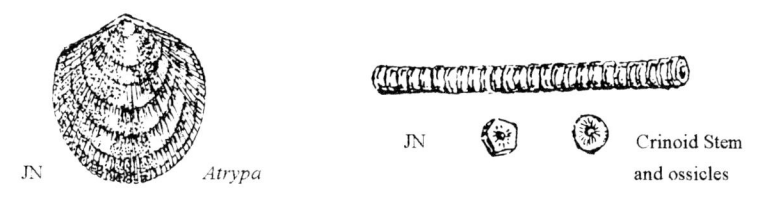

JN *Atrypa*   JN   Crinoid Stem and ossicles

For an impressive view, walk to the left (west) of the main quarry face and find a way over a gate or wire and then turn right into the higher workings, where loose rocks may be examined. You will see Wenlock Edge, a double escarpment, stretching away to the northeast and the mass of the Longmynd to the northwest. Keep your eyes open for large *Kirkidia*.

Kirkidium knighti - View Edge Quarry

Refreshments:
Bridges, *The Horseshoe Inn:* Lovely situation beside stream, between the Long Mynd and the Stiperstones. Wide range of real ales; lunches; non-smokers' lounge. Closed Monday midday.

Norbury, *The Sun Inn:* Fine ales; full range of bar snacks; lunches and evening meals (restaurant attached). Families welcome. (Closed Mondays).

Wentnor, *The Crown Inn:* Cosy village inn; real ales; huge range of meals (restaurant attached); booking advisable in high season. Accommodation.

Bishops Castle, *The Three Tuns Inn* (Salop St.); renowned home-brewed ales in a pub of great character; full range of snacks and meals midday and evenings; seven days a week.

THE HORSESHOE INN
at
Bridges
(Ratlinghope)

Unspoilt 16th century Inn set twixt the Longmynd and the Stiperstones

Adams Southwold Bitter, Adams Extra Shepherd Neame Spitfire, Guest Ales

Good Beer Guide
Listed 1993, 1994, 1995
Good Pub Guide 1995

Bar Meals (Lunchtimes only)

Tel: 01588 650260

The Ganges
Balti House
and Take-away

Hours of Opening: Open Seven Days a Week
Lunch 12 - 2pm   Evening 6pm - 12am

Balti House Meals;
Tandooris a speciality;
also English food;
bring your own wine

Partner:
M.A. Zafor

12 Market Square
Bishop's Castle
Shropshire. SY9 5BN
Tel: 01588 638543

THE CROWN INN

16th Century Country Inn & Restaurant

Fresh home cooked food

Traditional Ales, fine wines & malts

Accommodation

David & Jane Carr
The Crown Inn
Wentnor
Bishop's Castle
Shropshire
SY9 5EE
Tel: Linley (01588) 650613

"A rare example of a pub brewing it's own beer"

Extensive Menu
Fish ★ Vegetarian Dishes
Indian Specials ★ Steaks
Well stocked wine list

The Three Tuns Brewery
Bishop's Castle
(0588) 638797

Beer sold out in Polypins 4$^{1}/_{2}$ gallons. Ideal for parties

Brewery tours by appointment

ITINERARY H : THE ONNY VALLEY

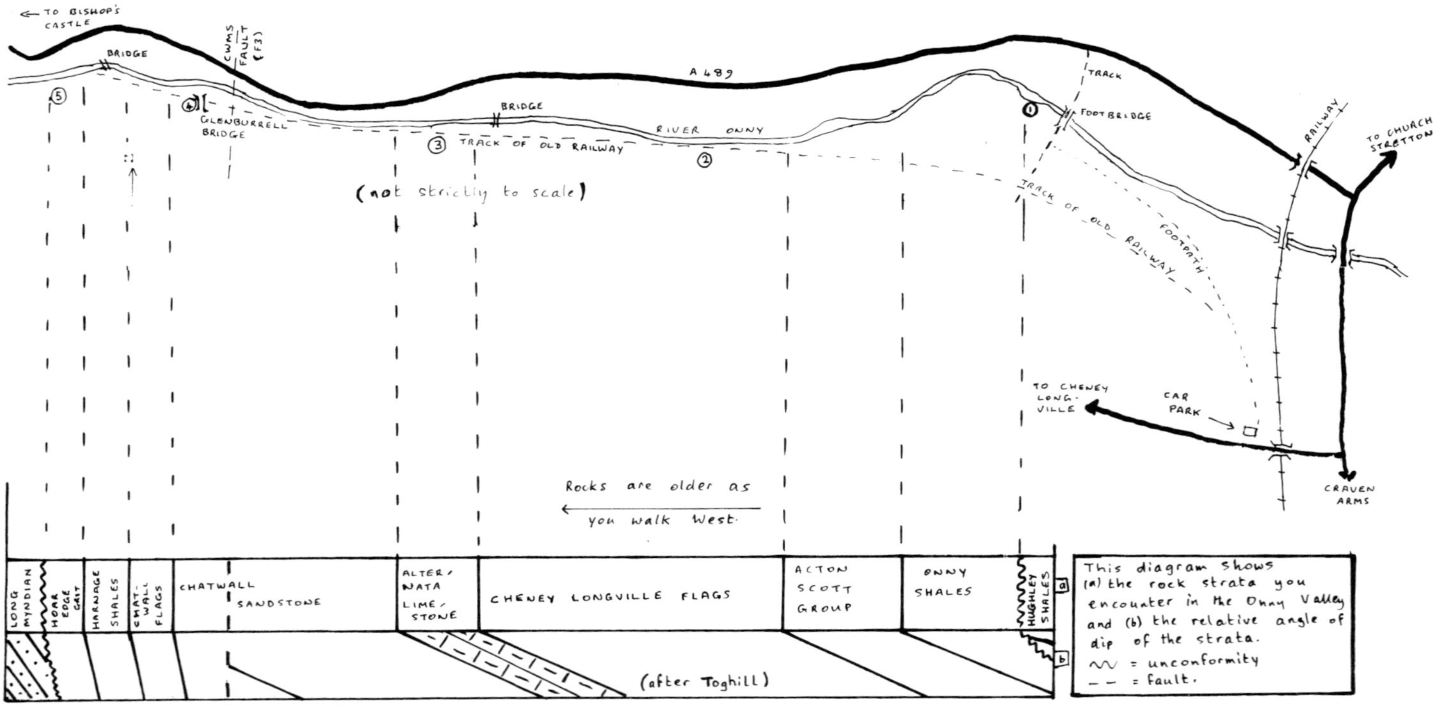

Itinerary E: THE ONNY VALLEY

This world famous section is of great importance to specialist geologists, but the casual visitor will also find much of interest. Features include the Ordovician/Silurian unconformity (E1), the Pre-Cambrian/Ordovician unconformity (E5) and a number of distinctive rock types, many of which are fossiliferous. It is a half day's excursion and a very pleasant walk along the riverbank, close by, or often on the line of the defunct Craven Arms to Bishops Castle railway.

Leave the A49 1 mile north of Craven Arms at the turn to Cheyney Longville. Over the railway bridge and immediately on your right is a signed car park (GR 430845). Turn in here. There is an interpretation panel with map, "GP" indicates certain geological points of interest; we have selected some of these and added others. Leave by the stile. Keep left through the cutting; then ascend the old railway embankment; continue through a gate and stile until you meet a sign post (here the embankment is right next to the stream). Go ahead, half right, across the hummocky field, essentially following the river, to (in quick succession) a stile, plank bridge, wooden steps, across a footpath and over another stile. You will see a footbridge over the river at this point; do not cross; but continue upstream for a further 360 yards.

E1. Ordovician/Silurian Unconformity (GR 427852)
Ordovician and Silurian Shales and trinucleid trilobites.
You are by the riverside, about 360 yards upstream from the footbridge. There is a wide beach (assuming the water level is not too high) on this southern bank, where you will find a great variety of stones, many of which have been washed down from the areas of the Stiperstones, Long Mynd, etc. Behind you is the disused railway. The unconformity is directly opposite, visible in the cliff section on the northern bank of the river. The lower (older) Ordovician Onny Shale beds dip (slope) southeast (ie down to your right) at $22°$ and upto 10 feet are visible above water level. The Silurian (younger) Hughley Shales above dip southeast at the slightly lesser angle of $18°$. This difference in angle indicates that, after the lower beds were laid down, there is a gap in the geological record; after this the Silurian rocks were deposited. However careful inspection is needed to pick out this angular unconformity. See photograph. The Onny Shales yield trinucleid trilobites and examples can usually be seen by examining pieces of the

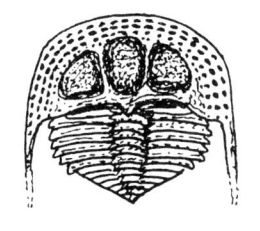

Trinucleid Trilobite

31

softer grey shales at water level and on the beach on the southern bank opposite the unconformity. Complete trilobites are unusual, but fragments are quite easily found. DO NOT CROSS TO THE NORTH BANK OF THE RIVER.

Onny Valley, Ordovician/Silurian Unconformity

E2. Old River Cliff (GR 421855)
Ordovician flagstones, brachiopods and Tentaculites sp..
This next locality is about 560 yards upstream (west). From E1 it's 100 yards to a stile; another 200 yards to the ruined railway bridge; and now walking along the railway track, a further 100 yards to the stile in the fence across the track; and 130 yards to the next fence and stile across the track. Beyond the stile you will see a very boggy area in front of the old cliff to your left; avoid this; continue past it and follow a track back left above the cliff where you will find the Cheyney Longville

Tentaculites

Flags exposed. In these sandstones and shales are bands containing brachiopods and also *Tentaculites sp* - small, straight, screw-like impressions, whose positive identity is uncertain; they may have been worms, or some kind of mollusc.

E3. Alternata Limestone (GR 418857)
Ordovician limestone and brachiopods.
Continue about 400 yards west from E2. You will see a footbridge over the river to your right. On your left, in the old railway cutting, are occasional exposures of a white/grey limestone, packed with the brachiopod *Heterorthis alternata*, preserved as a white film. Do not hammer the exposure; you may find specimens in the scree below, or loose boulders nearer the river.

Heterorthis alternata

E4. Chatwall Sandstone (Glenburrell Bridge) (GR 413860)
Ordovician sandstone and brachiopods.
From E3 continue some 450 yards along the track, passing through two gates. The track passes under the well-preserved Glenburrell Bridge. In the cutting, just prior to the bridge, the higher beds of the Chatwall Sandstone are exposed, dipping gently southeast; nearer the bridge the older Chatwall Sandstones are nearly vertical. This change in dip is due to a fault. There are exposures east and west of the bridge and the rocks contain brachiopods and occasional bivalves, gastropods and trilobites.

E5. Hoar Edge Grit/Longmyndian Unconformity (GR 412862)
Ordovician grit, brachiopods and Pre-Cambrian sandstone.
Follow the path for a further 250 yards west; you leave the railway track and a footbridge behind you on the right. A stile and gate give access to the old quarry on your left. The western end of the quarry, up a flight of steps, displays medium-grained purple/brown Pre-Cambrian Longmyndian sandstones dipping 70^0 southeast. Only about 6 feet are exposed and adjacent are the orange/brown Hoar Edge Grit beds of Ordovician age, also steeply dipping. Since they rest directly on the Longmyndian, the whole of the Cambrian is missing from the geological record, ie. there is a major unconformity. (A similar unconformity may be observed at Hope Bowdler - see Locality F5). The main face of the quarry consists of the Hoar Edge Grit; variable in grain size from fine sandstone to coarse grit; and containing brachiopods which are easily collected in the scree. At the extreme eastern side of the quarry the softer, younger, fine-grained,

greenish Harnage Shales can be seen resting on the Hoar Edge Grit - do not hammer. The Harnage Shales can also be seen in the bed of the river.

You may retrace your route to return to the car park, or go back to the track which uses the Glenburrell Bridge, cross over the bridge and follow the track south until you reach the road; turn left and the road will take you to the car park, via the village of Cheyney Longville.

Refreshments

Broome, *The Engine and Tender;* cosy old pub, serving excellent lunches and dinners; wide range of beers; caravan and campsite; caravan to let; near station and View Edge.

SRI LANKA (CEYLON)

Beaches : Mountains : Safari
Temples: Ancient Cities

For Information

* Flights
* Accommodation
* Activities
* Products
* Business

Contact:
Phillips Tutorials, Frogs Gutter,
Minsterley, Shropshire. SY5 0NL
Telephone: 01588 650335

"The Indian Ocean Paradise"

GEOLOGICAL DAYS

IN SHROPSHIRE

Let us introduce you to the Geology of Shropshire

* *Examine rocks and their relationship to the landscape.*

* *Collect and learn about a variety of fossils (including trilobites!).*

* *Explore the old lead/barytes mining area.*

* *Visit sites of past and present geological interest.*

* *Enjoy fresh air and wonderful scenery.*

Further details from:

The Crystal Man, Bridge Level, Pride Hill Centre, Shrewsbury. SY1 1BY.

Or telephone either 01691 650469 (anytime), or 01743 850318 (evenings).

LOCALITIES MAP

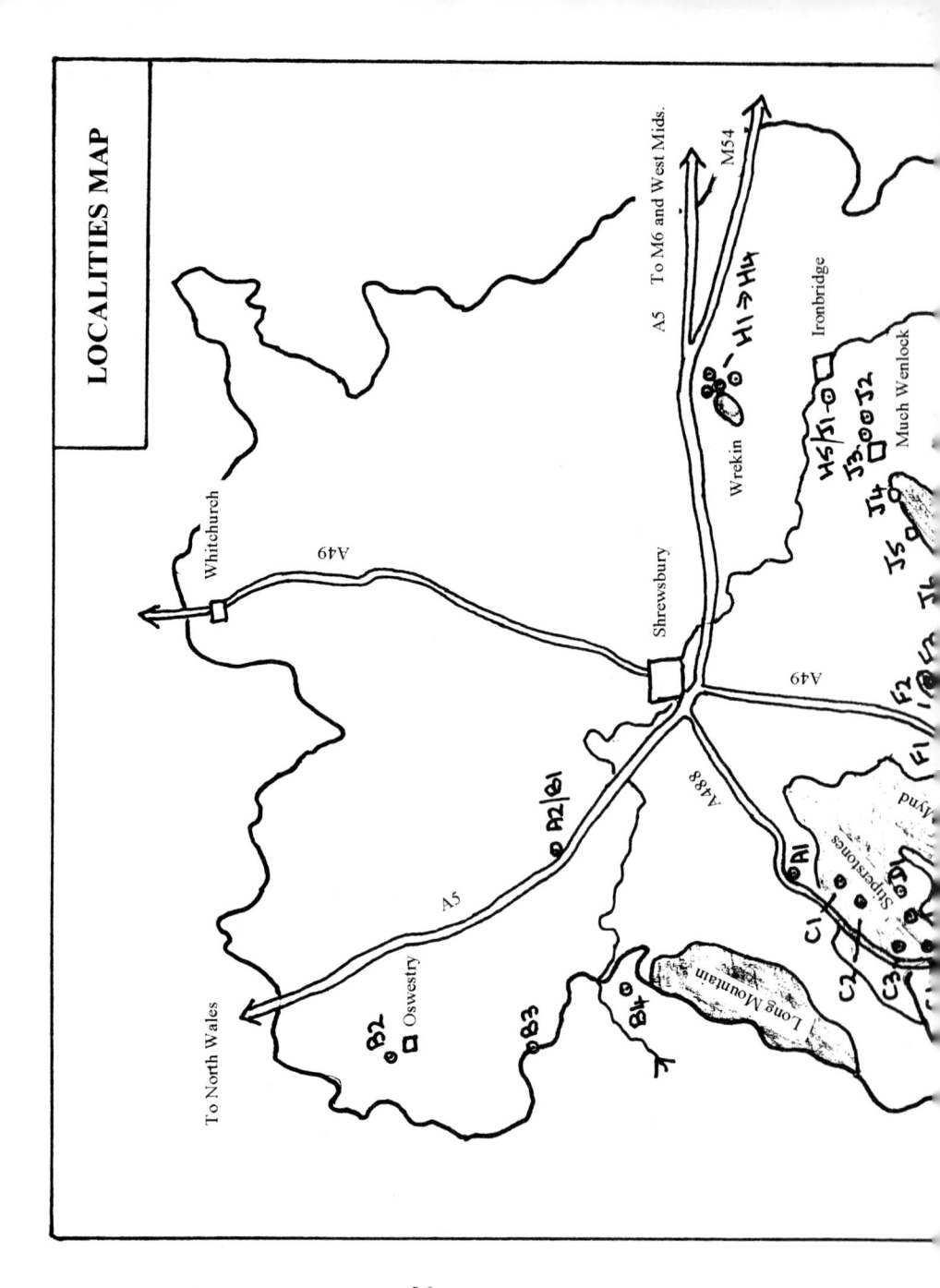

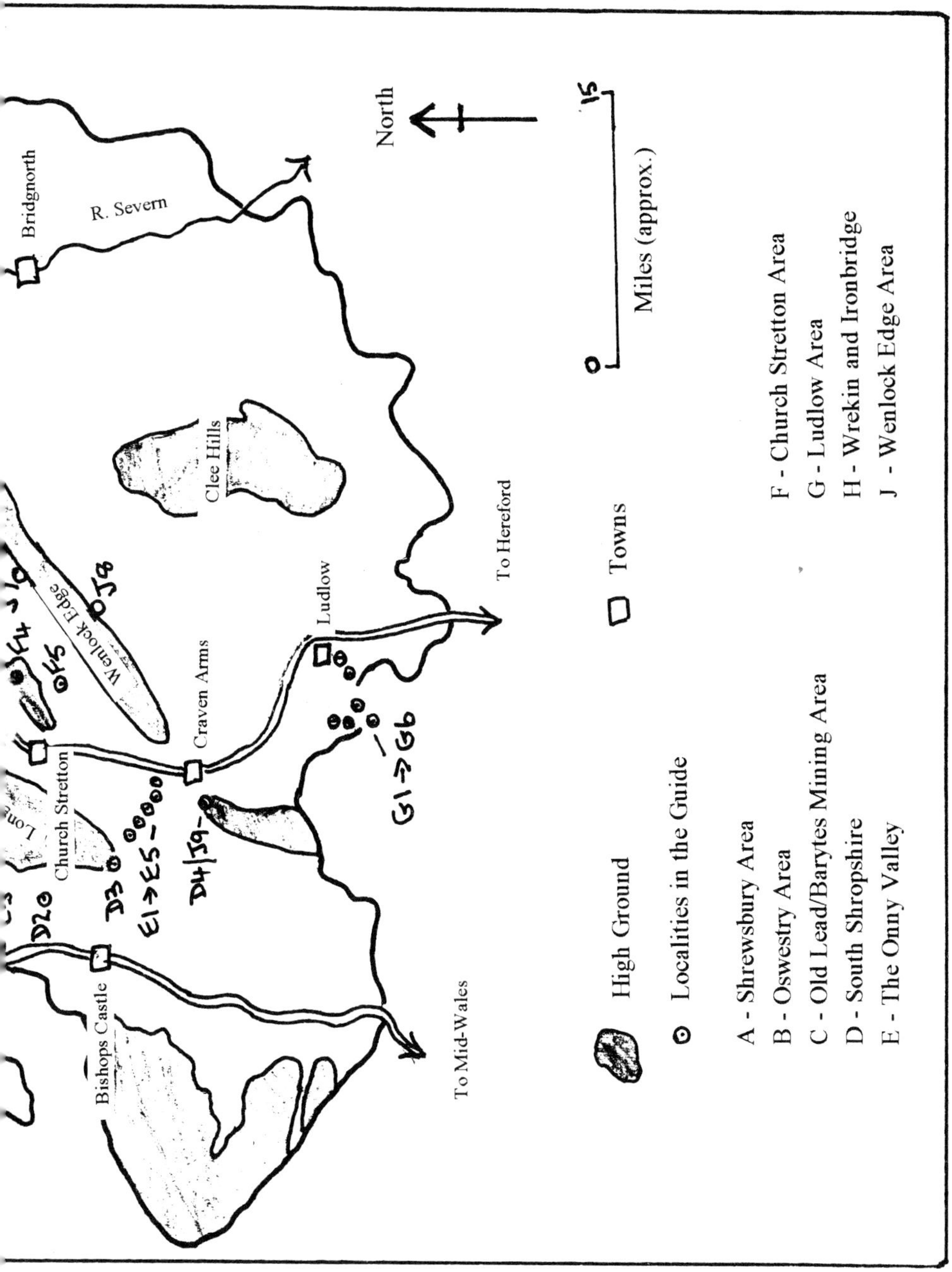
Bridgnorth
R. Severn
North
15
Miles (approx.)
0
Clee Hills
Wenlock Edge
J8
F5
J4
Ludlow
To Hereford
Craven Arms
Church Stretton
Bishops Castle
To Mid-Wales
D2
D3
E1→E5
D4/J9
G1→G6
High Ground
Localities in the Guide
Towns
A - Shrewsbury Area
B - Oswestry Area
C - Old Lead/Barytes Mining Area
D - South Shropshire
E - The Onny Valley
F - Church Stretton Area
G - Ludlow Area
H - Wrekin and Ironbridge
J - Wenlock Edge Area

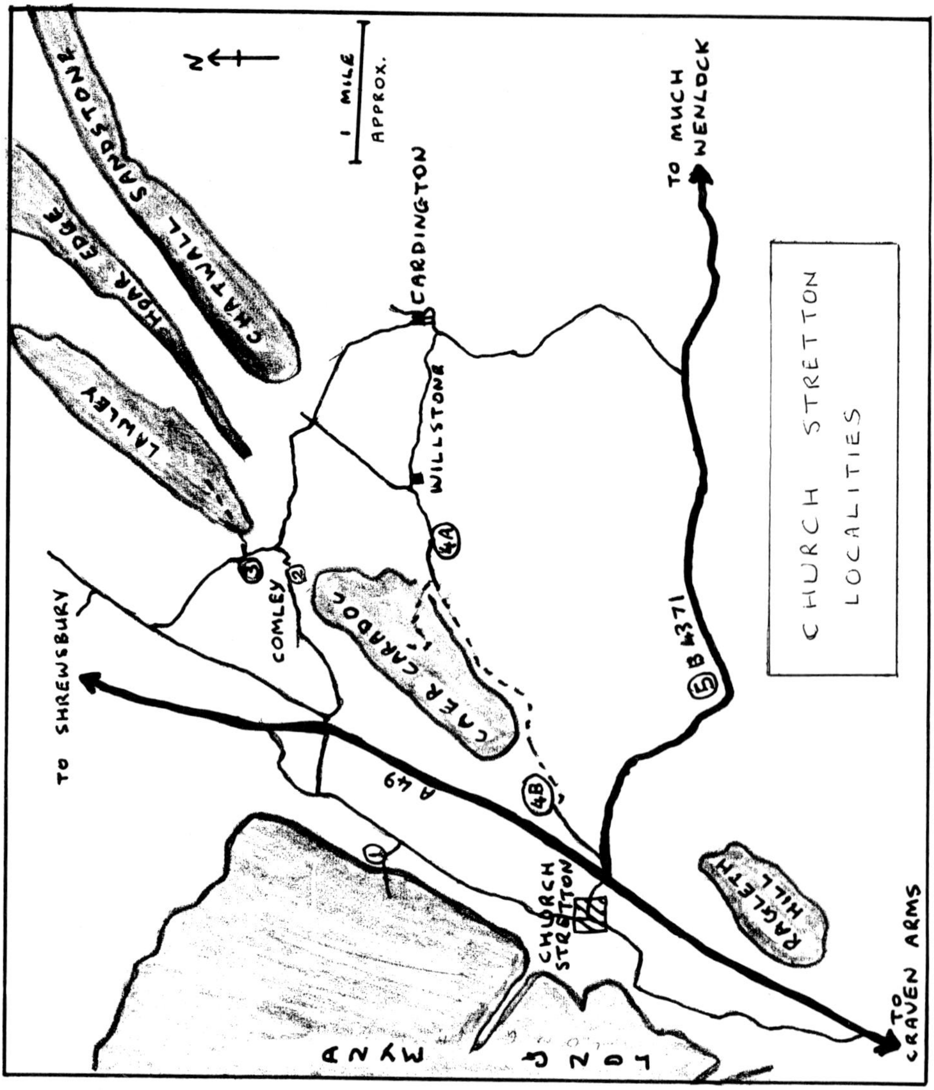
CHURCH STRETTON
LOCALITIES
N
1 MILE
APPROX.
CHATWALL SANDSTONE
HOAR EDGE
LAWLEY
CARDINGTON
WILLSTONE
4A
CAER CARADOC
COMLEY
2
3
TO SHREWSBURY
A49
4B
1
CHURCH STRETTON
B4371
5
TO MUCH WENLOCK
RAGLETH HILL
TO CRAVEN ARMS
LONG MYND

Itinerary F: CHURCH STRETTON AREA

As a centre for some of the loveliest hill country in Shropshire, the Stretton Valley has few equals. To the west runs the moorland plateau of the Long Mynd, cut into by enchanting valleys. To the east are the dramatic hills, Ragleth, Caer Caradoc and The Lawley, looking like extinct volcanoes. In the distance their line continues northeast to the Wrekin (see Intinerary H) and indeed, though they were probably not actual volcanoes, these conical hills run along an important fault line, where for hundreds of millions of years there was crustal instability, with earthquakes and volcanic activity.

These rocks are among the oldest in Shropshire and display great variety. It cannot be said that, for the complete beginner, the rocks are as striking or as obviously individual as some of the more recent ones; but a little inspection will reveal diversities of great interest.

We've designed a circular route to include the sites of geological interest. Note particularly, as you travel, how the landscape vividly reflects the geology, with tougher rocks forming hills and scarps and relatively less resistant rocks making vales.

F1. Buxton Quarry (GR 457958)
Pre-Cambrian Longmyndian shales and tuff.

Church Stretton is on the A49, 16 miles north of Ludlow and 13 miles south of Shrewsbury. At the traffic lights on the A49 turn west, though the centre of Church Stretton and then north along the B4370 to All Stretton. In the village turn left just before you reach the shop/post office, along the narrow lane signed "Village Hall"; pass a very narrow lane on your right and shortly after you'll see a small disused quarry on your right (in front of you is a "no through road"). You should be able to park just inside the entrance.

Looking northwards directly into the quarry you will see that the central area of the rock face consists of a 24ft wide band of a greenish grey rock, with small black spots and abundant thin quartz veins on the joints; and best described as a silicified dust tuff. This is known as the Buxton Rock and represents a spate of volcanic activity, between the laying down of the shales, which can be seen dipping steeply on either side of the quarry (the dip being approximately 70^0 west). The shales are Pre-Cambrian, Longmyndian; the younger, Burway Group, on the left and the older, Stretton Shale Group, on the right. A very fine near-vertical bedding plane can be seen in the shale on the right. Tap the Buxton Rock with a hammer and note how hard it is; by contrast the shales are much softer and give a different sound.

Return to the village of All Stretton and continue north along the B4370; take the first right; cross the A49 by turning left onto it and then immediately right; and at once right again at the junction in the minor road; continue for about 1 mile to Comley and you will see the next locality - a small overgrown quarry - on your right. About 100 yards past the quarry is a track leading off to the right and it is usually possible to park here without causing an obstruction.

F2. Comley Quarry (GR 484965)
Lower and Middle Cambrian sandstones, limestones and grits.
We include this locality because of its great importance in the history of geology in Shropshire. It is one of the few accessible exposures of rocks of Cambrian age in this part of Shropshire and is a reference point for correlating British Cambrian rocks with others worldwide. It was here, around 1890, that the great geologist Lapworth discovered the first Lower Cambrian fossil to be found in Britain. It was a trilobite, Olenellus callavei (now renamed Callavia callavei), from the Lower Comley Limestones. If you stand with your back to the road and look into the quarry you are facing south; on your right side are exposed approximately 14ft. of green glauconitic sandstone, with an easterly dip of $73°$; this is the Lower Comley Sandstone of Lower Cambrian age and forms the western face of the quarry. Lying on top of this steeply dipping sandstone, to the left of centre of the quarry are approximately 6ft. of the famous Lower Comley Limestones. The eastern face, or left side, of this small quarry is made up of younger, Middle Cambrian, sandstones and grits. Information is given on a board at the quarry entrance. Please do not hammer this important site; you will not find any trilobites! The rock may be examined by selecting pieces from the scree; the limestone can be distinguished from the sandstones by the fact that it is harder and more compact and by the "acid test". Add a drop of dilute hydrochloric acid (vinegar may do) to the limestone and it will fizz (give off carbon dioxide gas); there is no reaction with sandstone. Further detailed excavations and mapping here and in the local area were carried out by Cobbold in the 1920s.

NB. Unless you are feeling particularly energetic we think it unlikely that you will wish to attempt both the next two localities (The Lawley and Caer Caradoc) on the same day. They are each well worth a visit, but as far as geology is concerned the walk up Caer Caradoc probably has the edge.

F3, The Lawley (GR 494974)
Pre-Cambrian Uriconian amygdaloidal andesite.
Return to your car and continue in a north easterly direction; turn left at the first junction and after 400yds park opposite the entrance to a farm on the right. (On the left is a footpath sign "Ippikins Way"). Go right up the track, following the footpath sign, through a gate and continue on up. This develops into quite a steep climb, but the views are rewarding.
The hill consists of Uriconian Volcanics of Pre- Cambrian age; the rock which outcrops as you climb is andesite. This rock was deposited as a lava and in many samples numerous gas bubble holes are evident. These small cavities are frequently seen to be filled with pretty white or green crystals, formed later from percolating mineral solutions; and in this case the rock is described as being amygdaloidal. From the top observe the Wrekin to the north east, Caer Caradoc to the south west and to the south and east the 20 mile scarp of Wenlock Edge. Between the Lawley and Wenlock Edges are ridges of Hoar Edge Grit and Chatwall Sandstone; the vales between are in softer shales.

F4. Caer Caradoc (GR 477954)
Pre-Cambrian Uriconian rhyolite and vesicular andesite, Ordovician flagstones and grits, brachiopods, gastropods and polyzoa.
Return to your car and retrace your route to the junction north east of Comley; turn left and after about 1 mile turn right at the minor cross roads at the top of the hill; continue for a further mile to Willstone and then turn right up an unmade track towards Caer Caradoc.

An alternative approach to Caer Caradoc can be made from Church Stretton. Take the B4371 east from Church Stretton, after approximately four miles turn left at the signpost to Stoneacton and Cardington. (If you overshoot, continue past the Plough Inn and take the next left). After two miles (after the No Through Road on the left and just before Cardington), turn left, signed to Willstone and Caer Caradoc. After a further mile, where the road bends to the right, fork left up the unmade track signed to Caer Caradoc.

This track is rough but passable with care; after a ¼ mile there is a parking space on the right (GR 486952); leave your vehicle here. This is the site of a small disused quarry and the rock here is a purplish coloured sandstone or flagstone of Ordovician age and known as the Chatwall Flags. It contains fossil brachiopods and occasional small gastropods. Note that this rock forms a ridge running SW to NE - sandstone is a tough rock, resistant to weathering and erosion. Continue along the track on foot; it dips down into the next rock type, the older Harnage Shales - shales are less resistant rocks and are more easily eroded into hollows and vales. Pass through a gate across the track and then take the next gate, or stile, on the right where the path to the summit of Caer Caradoc starts.

NB. This point may also be reached along the track from the west - see end of itinerary.

Ahead of you the ground begins to rise as you traverse the next rock type, the Hoar Edge Grit - grits, like sandstones are tough rocks and form ridges and hills. Veer a little to the left and, just to the right of the fence alongside the track, you will see an exposure of a buff coloured coarse sandstone or grit; this is the Hoar Edge Grit. There are plenty of fossils here, although many are not particularly well preserved (a coarse-grained rock preserves detail less readily than a fine-grained shale). Brachiopods (*Rhynchonellids*, curved with prominent ribbing and *Orthids*, flatter and smoother) are the most common fossil types, although you may find good examples of polyzoa, both stick and lace forms.

JN   "Stick" Polyzoan   "Lace" Polyzoan

Return to the path and commence your climb. Caer Caradoc is ahead and a little to your right (NE) you will see The Lawley. To the right (east) of The Lawley you will see a series of vales and ridges. The steep face of The Lawley (Pre-Cambrian Uriconian) dips into a vale of soft Cambrian rock; the land rises to the wooded ridge of Hoar Edge Grit; dips into a valley of Harnage Shales and then up again onto a ridge of Chatwall Sandstone. The path crosses a fence at a stile and begins to climb more steeply. Further on, on either side of the path you will notice boulders of a much-weathered brownish pink rock and above you the same rock forms a near vertical cliff. This is rhyolite and very old (Uriconian); some samples exhibit flow-banding and appear "streaked". Where the path bears to the right before the final leg to the summit there is a flattish area and here there are small low exposures of andesite, a dark grey/black rock, many samples of which are vesicular, ie show gas bubble holes. This is the same rock type as appears on The Lawley, but here on Caer Caradoc it is unusual to find the holes infilled.

The view from the summit is well worth the climb. To the west the vale of Church Stretton, the dissected plateau of the Long Mynd and beyond this, on a clear day, the rocky tors of The Stiperstones are just visible. To the east are the already mentioned ridges of Hoar Edge Grit and Chatwall Sandstone and further

SOUTH SHROPSHIRE PERSPECTIVE.

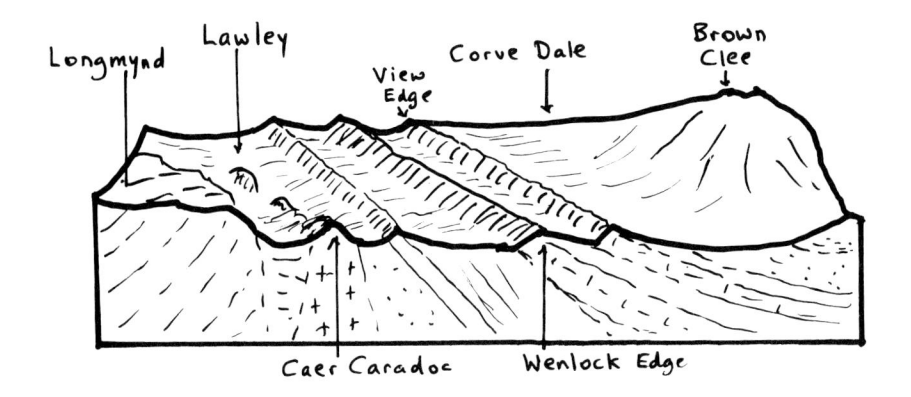

Diagram labelled: Longmynd, Lawley, View Edge, Corve Dale, Brown Clee, Caer Caradoc, Wenlock Edge.

eastwards the long wooded escarpment of Wenlock Edge. In the far south east rise the Clee Hills. Immediately north is the whale back shape of The Lawley and in the distance, on the same line. The Wrekin; with the steam from the Ironbridge power station nearby.

Caer Caradoc is an ancient hill fort and at least three concentric defence mounds are readily identified.

There is a third route to Caer Caradoc, but this does involve a longer walk. Take the B4371 east from Church Stretton; turn left into Watling Street. Ignore the track to the left. Where Helmeth Road goes right, take the single track road to the left. At the entrance (cattle grid) to New House Farm, you'll see a footpath sign pointing right (not the one 20 yards earlier). Parking here is not easy. The footpath follows the edge of the field into a wood (parallel to a track); fork left over a bridge. Shortly, the footpath forms a broad track; continue, passing two "Ippikins Way" signs on your left and the ruined Cwm Cottage on your right. Take the next footpath left, through a gate or over the stile and you are on the path to the summit of Caer Caradoc. Continue with earlier directions.

If you are returning via Church Stretton, then we suggest you take a look at:

F5. **The Hope Bowdler Unconformity** (GR 474924)
Join the B4371, Much Wenlock to Church Stretton road at either Wall Bank or Wall Under Heywood and turn west. The village of Hope Bowdler is 1½ miles

southeast of Church Stretton; pass through the village and as the road climbs up the hill you will see a layby on your right; park here. The site is on the bank on the opposite side of the road and shows Ordovician Harnage Shales resting unconformably on Pre-Cambrian Uriconian volcanics. There is an explanatory sign.

Refreshments:

Wall, *The Plough Inn*. Traditional country pub with large restaurant attached; extensive menu; excellent beer and food; well-priced.

Little Stretton, *The Ragleth Inn*. Charming site, near Ashes Hollow. Good beers and good food seven days a week; accommodation; beer garden. Warm Olde World atmosphere.

Church Stretton, *The Acorn Restaurant* (26 Sandford Ave.). Wholefood cafe and coffee shop serving beverages, snacks, cakes, light meals. Tea garden. Egon Ronay recommended.

Church Stretton, *The Book Cellar* (61 High Street). Open Monday to Saturday. Coffee, snacks, lunches and afternoon teas. Secondhand and first edition books for sale.

Church Stretton, *The Coffee Shop* (The Square). Open all day, seven days a week for beverages, snacks and full meals. In the centre of the town.

THE COFFEE SHOP

The Square, Church Stretton

Open all day, seven days a week
for
Beverages, Snacks, and Full Meals

In the centre of the town

LUDLOW

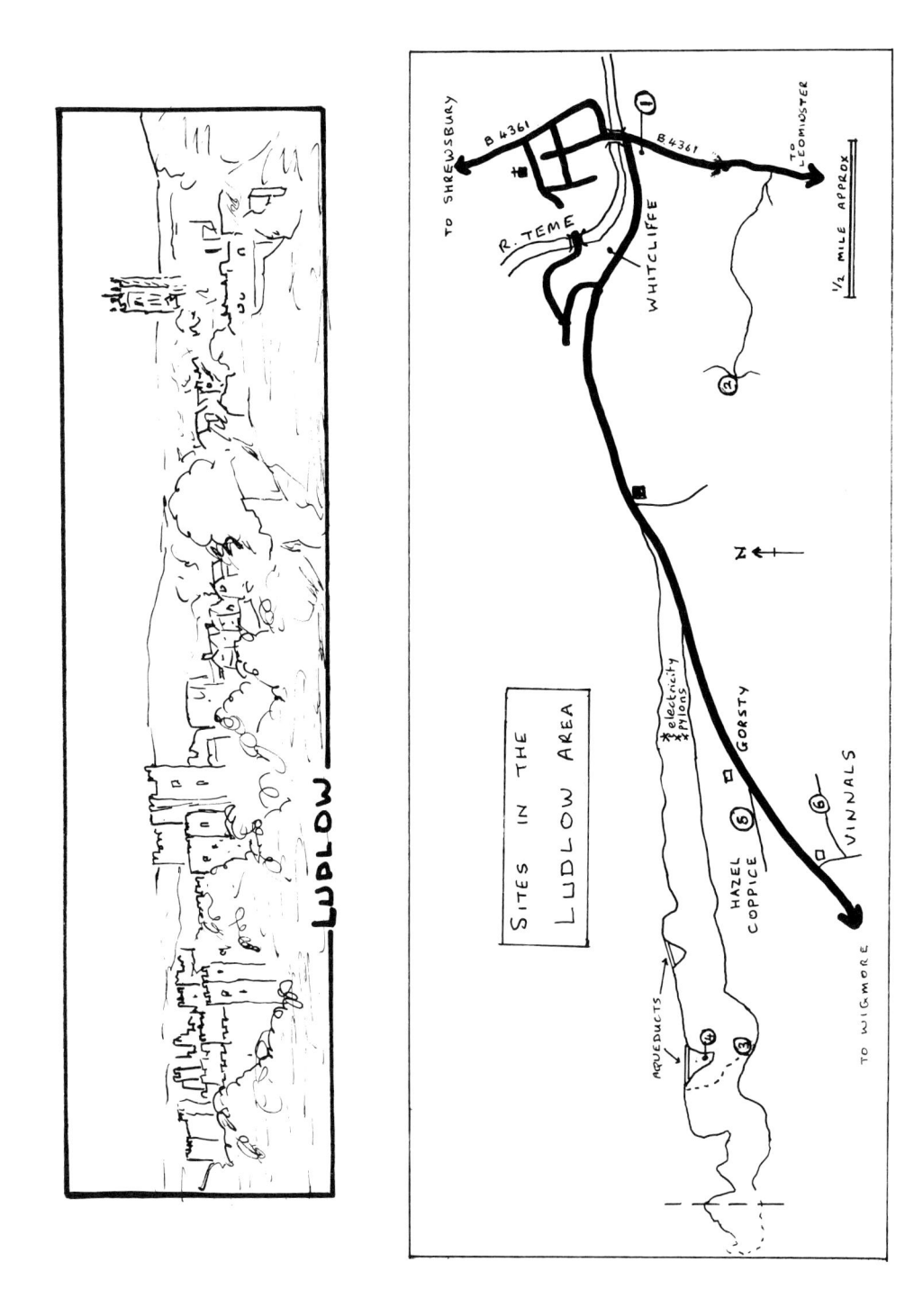

SITES IN THE
LUDLOW AREA

Itinerary G: LUDLOW AREA

The rocks and fossils of the Ludlow area are world famous. A fascinating new display "Reading the Rocks" is now open at Ludlow Museum (Castle Street). It is suitable for all ages with 'hands on' opportunities for visitors; also the history and growth of the town. There is a Mortimer Forest Geology Trail by Andrew Jenkinson, published by Scenesetters, which we highly recommend. For the following itinerary we have chosen sites where the geology and fossils are self-evident and accessible.

G1. **Ludlow Bone Bed** (GR 513742)
Silurian/Devonian fish remains.
Leave Ludlow on the old trunk road, that is down Broad Street or Old Street and over the Dinham Bridge. Immediately there is a right turn by the Charlton Arms pub, to the Whitcliffe. Turn here and park as safely as you can. Alternatively park in the town and stroll out, it's no distance.
On the southern corner of the lane, opposite the pub, you will see the bone bed, as a deep notch, chiselled into the rock face by generations of geologists. Don't hammer here, but you may find fragments. For details of its appearance see "Fish" in the fossil Glossary. Walk just along the Whitcliffe Lane and, on the north side, you will find similar chiselled notches, where it may be possible to collect a sample. A piece 1 inch across is reasonable. It is not impressive, perhaps, in itself; but its historical significance is awesome.

Within the Silurian rocks west of Ludlow you will find a variety of fossils - including brachiopods, bivalves, cephalopods, early gastropods, graptolites and trilobites (but no ammonites or dinosaurs!) - an extraordinary range of life forms; though not of course at every site.

G2. **Tilhill Forestry Quarry - Whitcliffe Beds** (GR 499736)
Silurian mudstones and shales, bivalves, brachiopods, cephalopods, gastropods and worms.
This is an excellent quarry for finding cephalopods. It is private: permission to enter may be obtained by telephoning the Manager, Tilhill Forestry, 01694 781511.
Take the B4361 south from Ludlow. You are looking for the entrance to a house called Mabbert's Horn; it is on your right about ½ mile from Dinham Bridge. Turn up the drive, bear right, and after 100 yards or so, you will see a gate and stile on the right, leading towards the wooded area (if you come to a dead end, at a

house, you have gone too far). Park off the track opposite the gate. Go over the stile and follow the track, leaving some pools, first on your right and later on your left. The obvious track takes you, in all, rather less than a mile through the attractive belt of woodland. You will eventually come to a T-junction (just after crossing a little brook); turn right and the quarry is on your left. It has a wonderful population of cephalopods - probably the best chance in Shropshire of seeing them. You will find long sections, both smooth and ribbed, and round septa, of a number of varieties. eg. *Orthoceras* and *Dawsonoceras*. Additionally there is the distinctive bivalve *Fuchsella amygdalina*; worms (*Serpulites sp.*), upto 18 inches long!; and even the occasional gastropod. The trackside screes contain abundant brachiopods. eg. *Camarotoechia*, *Isorthis* and *Protochonetes*.

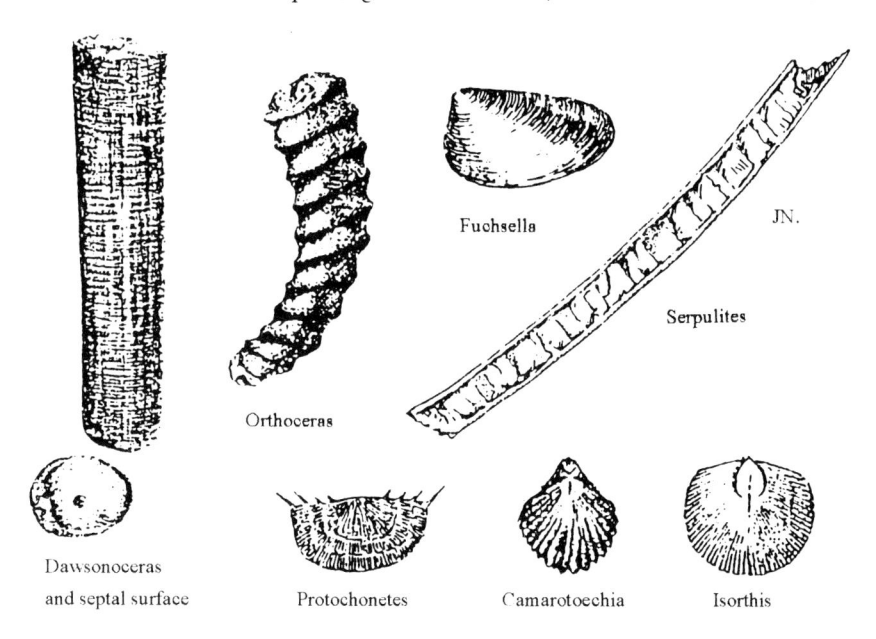

Fuchsella

JN.

Serpulites

Orthoceras

Dawsonoceras
and septal surface

Protochonetes

Camarotoechia

Isorthis

Return to the Whitcliffe (which provides a stunning view of Ludlow town against the Clee Hills) and take the Wigmore road west. After about ¾ mile, the Forestry Commission main entrance and car park is on your left (GR 494742). There are several nature trails here and the Mortimer Forest is a lovely place to spend an hour or two. Note the closing times and if necessary park outside the gate.

G3. **Mortimer Forest, Aymestry Limestone**: (GR 464736)
Brachiopods.
Approximately ½ mile west of the Forestry Commission car park you will see

an entrance to the forest on the right (GR 488739). You can park carefully here and enter the forest around the padlocked gate. About 1½ miles along this track - see map - where an obvious footpath bears down right, there is an excellent exposure of the Aymestry Limestone (Bringewood Beds), where the brachiopod *Atrypa* is especially abundant.

G4. **Mortimer Forest, Leintwardine Beds**: (GR 465738)
Silurian mudstones, shales, brachiopods, cephalopods, trilobites and worms.
From the previous site, follow the steep footpath downhill and you will meet a track. Turn right (ie east); you follow open ground on the edge of the forest and will see an aqueduct (it caries water from the Elan Valley in Wales to Birmingham - a splendid piece of Victorian engineering). The track, unable to cross the aqueduct, loops around a gully. Down on your left near the stream is an excellent exposure of the Leintwardine Beds, with a rich fauna and easy collecting. Fossils include the small brachiopod *Dayia navicula* and the larger *Salopina lunata*, trilobites, cephalopods and worms.

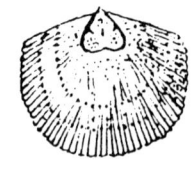

Dayia   Salopina   JN.

You may return the same way or continue east along the track until you reach the footpath alongside the electricity pylons. There are frequent dumps of rock waste on and alongside the track where fossils (brachiopods, trilobites, etc.) may be found. There are also much longer routes back to Hazel Coppice and to the Forestry Commission car park (see map).

G5. **Gorsty**: (GR 477736)
Silurian mudstones and shales, graptolites.
Further west along the road to Wigmore, about 1¼ miles from the Forestry car park, look for a major track on the right - it may have a locked gate. There is a signboard, Hazel Coppice. Park so as not to obstruct the entrance Walk a little way along the track and you will find the shale bedding planes contain graptolites (*Monograptus*). This is an unusually good location for collecting these enigmatic creatures.

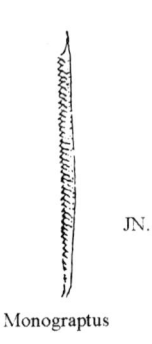

JN.

Monograptus

G6. **Vinnals**: (GR 474732)
Silurian mudstones and shales, cephalopods, graptolites and trilobites.
Travel a further 600 yards to Vinnals car park, on your left. Go out of the car park 250m (sign: Mortimer Forest Geological Trail Stop 4) up (south) the main track.

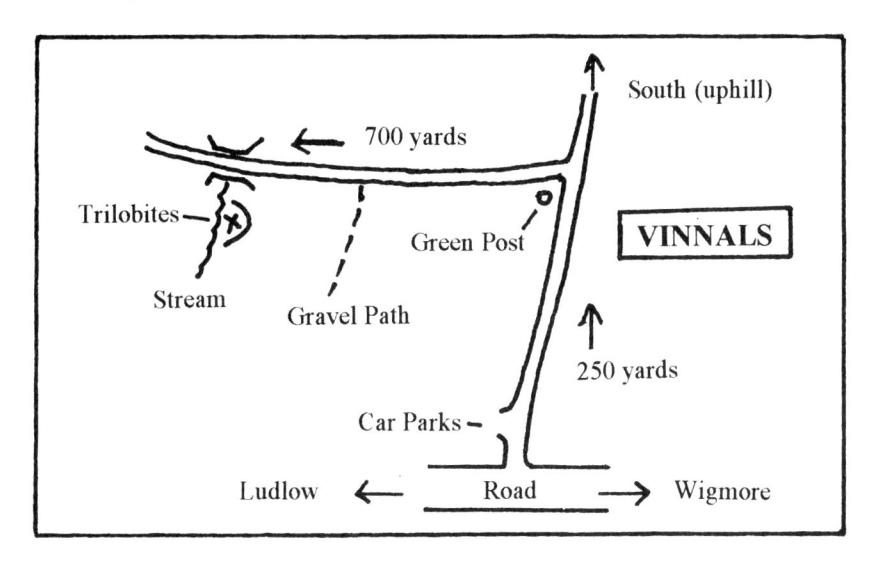

```
                                                  ↑
                                              South (uphill)
            ___    ← 700 yards
       ___/   _____
                                           |
   Trilobites —⟨✗⟩                    o    |
                  \          Green Post\   |    ┌─────────┐
       Stream      \                        |    │ VINNALS │
              Gravel Path                   |    └─────────┘
                                            |
                                            |    ↑
                                            |  250 yards
                                            |
                            Car Parks —  ___|
                                        |
         Ludlow  ←                Road              →  Wigmore
```

At the junction turn left (green post). Walk ½ mile passing a gravel path on your left, and you will come to a small stream - sign Mortimer Forest Geological Trail Stop 4. In the rocks by the stream on the left of your path you will see the evidence of fossil hunters. Well preserved trilobites (*Dalmanites*) are plentiful at this locality and you should be able to find a number of specimens - individual heads and tails, rarely bodies as their segments have usually separated and dispersed. You may find cephalopods and graptolites are also present but scarce.

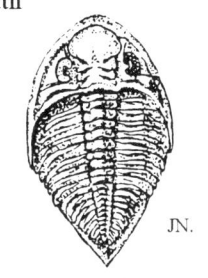

Dalmanites

Between the Vinnals car park and Elton, on the road to Wigmore, there are several small quarries/exposures on the roadside which are worth examination; the more westerly ones being in Wenlock limestone. Also, at GR 456713, just along the turn off to Burrington, you can obtain an excellent view of the eroded Ludlow anticline. See the Mortimer Forest Trail Guide for further details.

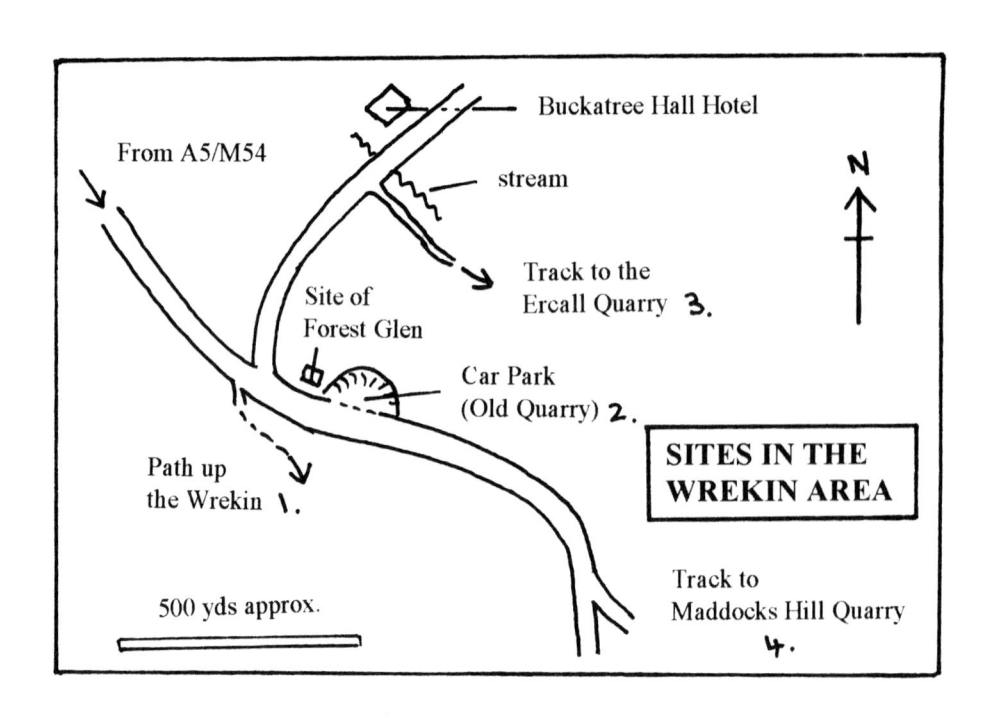

SITES IN THE WREKIN AREA

- From A5/M54
- Buckatree Hall Hotel
- stream
- Site of Forest Glen
- Track to the Ercall Quarry **3.**
- Car Park (Old Quarry) **2.**
- Path up the Wrekin **1.**
- Track to Maddocks Hill Quarry **4.**
- 500 yds approx.
- N

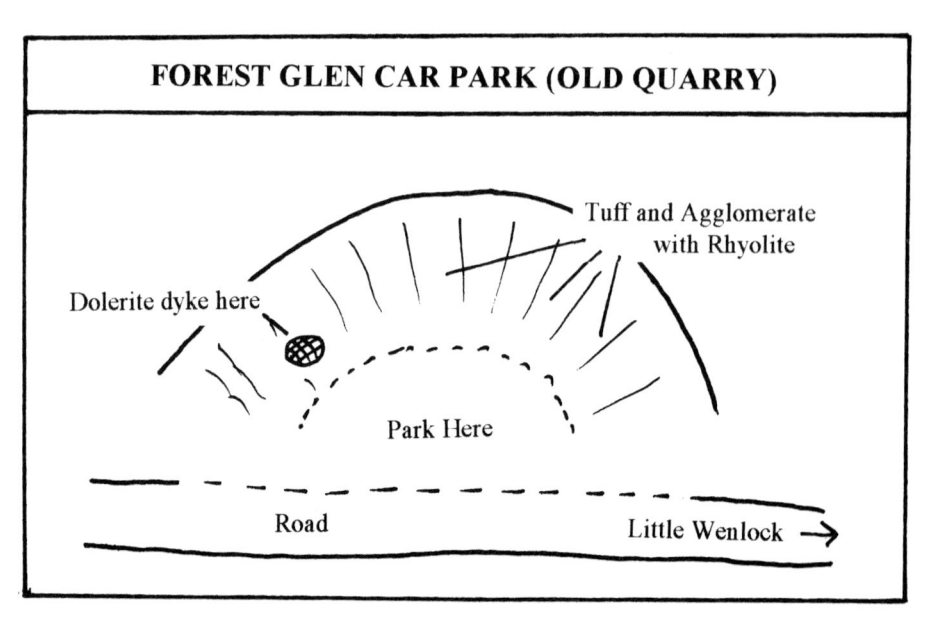

FOREST GLEN CAR PARK (OLD QUARRY)

- Dolerite dyke here
- Tuff and Agglomerate with Rhyolite
- Park Here
- Road
- Little Wenlock →

50

Itinerary H: THE WREKIN and IRONBRIDGE GORGE

The Wrekin is a well-known, prominent hill in the flat northern Shropshire plain, about 9 miles east of Shrewsbury.

While the geology is complicated, it is difficult to resist visiting the Wrekin for its splendid views - the Shropshire plain to the north, the Shropshire Hills to the south and, to the east, the industrial Midlands. The climb is not too arduous, taking the track from near the car park of the, now demolished, Forest Glen Pavilion. Nearby are several quarries that are certainly worth a visit; the rocks are of great interest scientifically.

Turn off the A5 where it enters the M54. Signposted (from Shrewsbury direction) B5061 Wellington, about 9 miles east of Shrewsbury. At the end of the short slip road turn south, ie right, away from Wellington, signposted "The Wrekin and Little Wenlock". Follow the road for ¾ mile to a T junction, where you turn right, signed "Little Wenlock, Huntsman Inn", and the Forest Glen car park is immediately on your left.

H1. **The Wrekin**
The track leads up the hill from the T junction behind you (see sketch map).

H2. **Forest Glen Car Park** (GR 638093)
Uriconian rhyolite tuff and agglomerate and dolerite.
There are three features of geological interest visible here. (see sketch diagram).
(a) Most of the rock you can see as you face into the quarry is either a tuff or an agglomerate. The tuff is a hard compacted volcanic ash, with small red, green and purple particles, representing ash blown out from volcanic vents more than 600 million years ago; it is a very striking rock. The agglomerate is similar, but is made of larger particles, some more than 6 inches in length.
(b) Rhyolite - this is another igneous rock, formed of surface lavas. It is grey/brown and very fine-grained. Some samples show banding, indicating the flow of the lava before solidification.
(c) Dolerite dyke. A dyke is formed of igneous rock which cooled below ground level. It was squeezed through weaknesses in surrounding rocks. It is therefore younger than the surrounding rock. You can see the dyke on the left of the quarry, in greenish dolerite.

H3. **Ercall Quarry** (GR 643096)
Cambrian conglomerate and quartzite.
Walk out of the Forest Glen car park to the right and follow the road round to

THE ERCALL QUARRIES

- Wrekin Quartzite with Bedding Planes
- Wrekin Conglomerate
- stream
- earth ramp
- Road
- Forest Glen →
- Buckatree Hall Hotel
- ═══ track
- --- paths

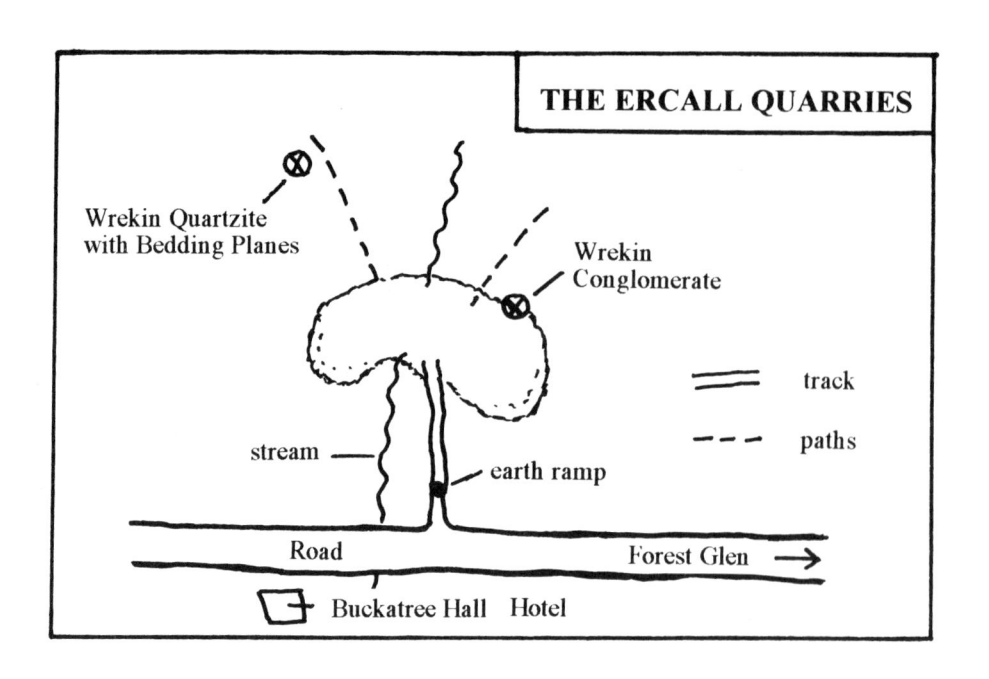

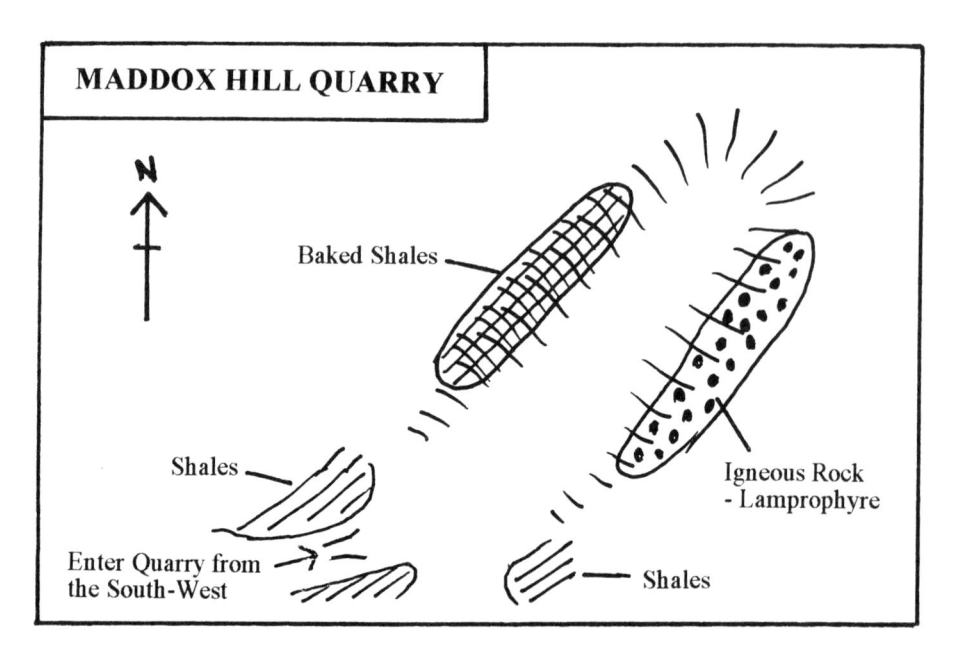

MADDOX HILL QUARRY

N ↑

- Baked Shales
- Igneous Rock - Lamprophyre
- Shales
- Enter Quarry from the South-West
- Shales

the right (see sketch map). Continue for about 500yds; the entrance to the quarry is on your right, about 50yds before the Buckatree Hall Hotel. Walk up the track and through the trees into the quarry (the earth ramp is to prevent vehicles entering). Continue along the track until you reach a junction of many paths. To the right of the right-hand path (see diagram) you will find the most unusual rock, Wrekin Conglomerate. This is a shallow water rock, composed of fragments of the Wrekin itself which were broken away by the shallow Cambrian sea. Some of the pebbles are really very large and the rock resembles poorly set concrete.

If you now walk along the left-hand track (see diagram), you will find on your left a large quarry face in Wrekin Quartzite (a very tough sandstone). The feature is the striking bedding planes - that is, the obvious steep slope representing the layers of rock as it was laid down on the ocean floor and later tilted by earth movements. Return to the Forest Glen car park.

H4. **Maddox Hill Quarry** (GR 647088)
Ordovician shales and igneous intrusion.

Turn left out of the Forest Glen car park. First left (after 500 yards) is a track to the quarry (OK for cars). Proceed along the track for 250 yards until you see a gate on the left behind a number of very large boulders. Park and enter the quarry via the gate, ie. from the south west.

This quarry illustrates an igneous intrusion, in this case known as a sill, in sedimentary shales. The igneous rock is described as a lamprophyre; it is of striking appearance, being medium-grained with red and black particles. Most of it has been removed for use as roadstone and the quarry closed a few years ago. You will find plenty of lumps of this rock on the quarry floor and, in situ, on the righthand (eastern) wall. The molten rock (magma) was intruded, during late Ordovician times, into the older surrounding rocks, which are of early Ordovician age and are known as the Shineton Shales.

The shales are grey/green when fresh, but a rusty brown when weathered and have yielded a few dendroid graptolites. Their dip is almost vertical in places. The fierce heat of the intrusion has baked the shales, which become harder the nearer they are to the igneous rock (ie increasingly metamorphosed). On the left-hand (western) wall of the quarry you will see the shales have been transformed into a tough splintery rock.

To Ironbridge:
Return down the track to the road and turn left towards Little Wenlock. Take

the second right, at the Huntsman Inn. At the Y-junction, signed Coalbrookdale and Ironbridge, turn right; cross over the new by-pass, down into Coalbrookdale (fine views), under the bridge and at the T-junction and main road turn right to Ironbridge. At the next T-junction (mini roundabout) turn left towards the centre of Ironbridge. There is a large car park almost immediately on your right, at the Museum of the River; there are also car parks about ½ mile further on near the iron bridge and, continuing to the roundabout and turning right, on the Coalport road.

There are a number of cafes and pubs for refreshment.

Visit:
The Jackfield Tile Museum; "The Great Rock Sandwich" is an exciting gallery, dedicated to the geological and mining history of the East Shropshire Coalfield.

H5. **Pattinswood Quarry** (GR 662034)
Wenlock Limestone, brachiopods, corals, crinoids, gastropods and polyzoa.

Wenlock Limestone Fossils

Note this is a tough walk. Cross over the iron bridge and turn right. You will in fact be following one of the signed "walks" and can buy a guide at the toll house; but it's singularly unhelpful.

Follow the track of the old railway line, under the bridge; continue for about ½mile and then, just before reaching the foot of the first of the power station cooling towers, turn sharp left back uphill. At the clearing, with a bench seat, continue straight uphill leaving the bench on your right. Climb up an exhausting flight of steps and almost at the top the path and handrail turn right; on this corner you will see where people have entered the quarry. Here you will find fossilised all the teeming life of the Wenlock Limestone (reef and massive); brachiopods, corals, crinoids, gastropods, polyzoa, etc. It is well worth spending an hour here. We do not advise you to leave the quarry other than by the way you came!

The Iron Bridge

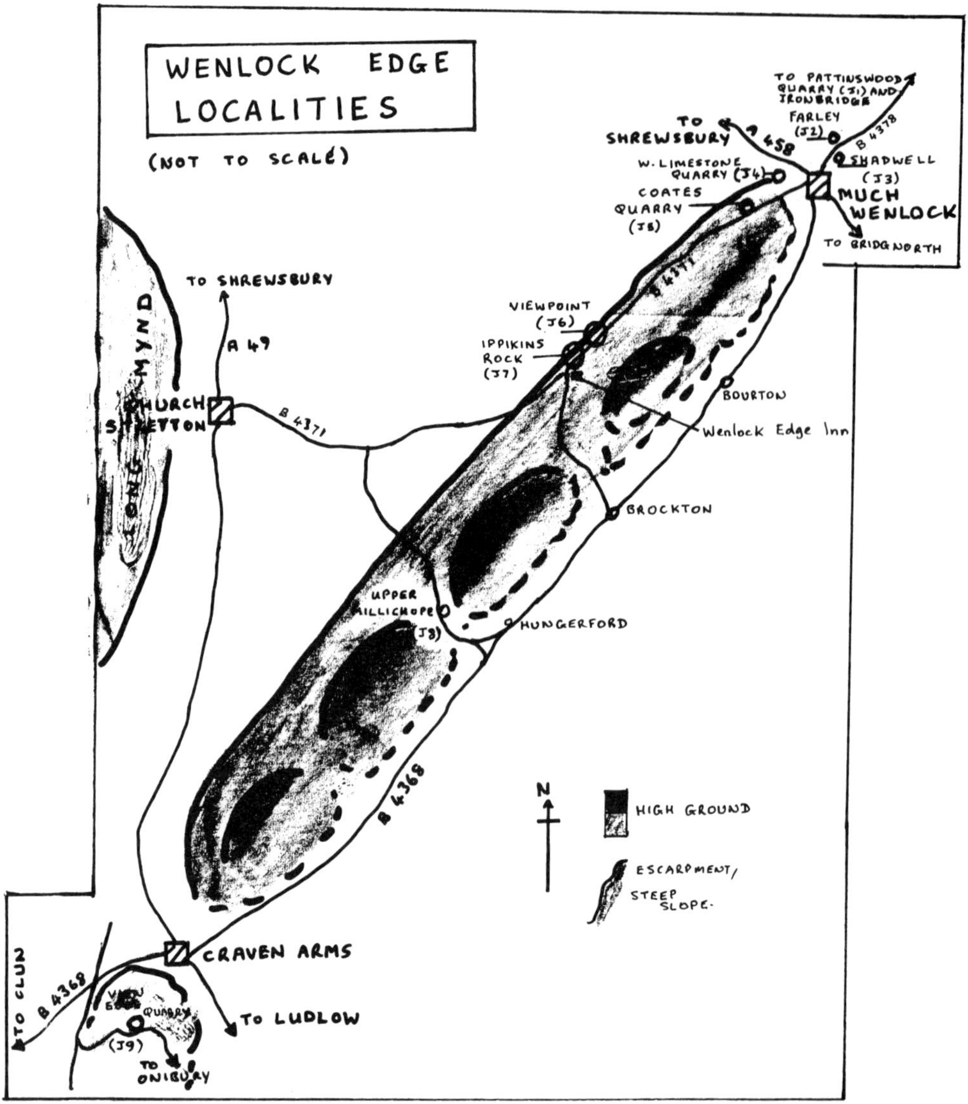

WENLOCK EDGE
LOCALITIES
(NOT TO SCALE)
TO PATTINSWOOD QUARRY (J1) AND IRONBRIDGE
FARLEY (J2)
B 4378
TO SHREWSBURY
A 458
W. LIMESTONE QUARRY (J4)
SHADWELL (J3)
COATES QUARRY (J5)
MUCH WENLOCK
TO BRIDGNORTH
B 4371
TO SHREWSBURY
A 49
LONG MYND
VIEWPOINT (J6)
IPPIKINS ROCK (J7)
BOURTON
Wenlock Edge Inn
CHURCH STRETTON
B 4371
BROCKTON
UPPER MILLICHOPE (J8)
HUNGERFORD
B 4368
N
HIGH GROUND
ESCARPMENT/ STEEP SLOPE.
CRAVEN ARMS
TO CLUN
B 4368
QUARRY
(J9)
TO ONIBURY
TO LUDLOW

Itinerary J: WENLOCK EDGE

Wenlock edge is a striking feature, a straight double escarpment, running some 20 miles from Ironbridge to Craven Arms. See photograph on page 11. It offers splendid views; and its geology is world famous. Although there are a number of large quarries they are not usually open to the public. The front (lower) escarpment is of Wenlock Limestone, which is very variable in nature, appearing as bedded, nodular, crystalline or crinoidal "massive" limestones; and the northerly part of the outcrop, ie. north east from Easthope Wood (GR 562957), contains numerous areas of large fossilised reef formations, ie. "reef" limestone. In the warm shallow seas of that time life was abundant. Fossils include corals, brachiopods, crinoids, gastropods, trilobites, bryozoa, algae, etc. - often in great numbers.

Aymestry Limestone forms the higher but dissected escarpment. It is a somewhat harder rock; a massive limstone, also containing abundant fossil life, including both circular and star-shaped crinoid ossicles and many brachiopods, including the giant *Kirkidium*.

J1. **Pattinswood Quarry**, near Ironbridge, (GR 662034).
See Itinerary H: The Wrekin and Ironbridge Gorge, Locality H5.

J2. **Farley Quarry** (GR 630015)
This is a very large privately owned quarry in the Wenlock Limestone, on the west side of the A4169 about 1½ miles north of Much Wenlock. If you can track down the present owners and obtain permission to enter, it should be well worth a visit.

J3. **Shadwell Quarry** (GR 625010)
This is another large working quarry in the Wenlock Limestone, on the east side of the A4169 1 mile north of Much Wenlock and it is possible that you may be allowed to visit. You should request permission at the entrance.

Continue west along the A4169 to the town of Much Wenlock, with its Abbey.

Refreshments:
Much Wenlock, Ye Olde Courtyard (High St.). Cafe and restaurant with a very wide range of delicious home-cooked food, competitively priced. In the heart of Much Wenlock.

Visit:
Much Wenlock Museum; geology display "On Wenlock Edge"; fascinating model of a "living" Silurian Sea; relationship between limestone and wildlife; fossils to look at and touch; town history.

Leave the town by the A458 and then B4371, heading towards Church Stretton. Shortly after leaving the A458 and turning onto the B4371 you will see a National Trust car park on the right (GR 613997); park here.

J4. **Wenlock Limestone Quarry** (GR 613999)
Brachiopods, corals, crinoids, polyzoa.
Leave the car park on foot via the small wooden gate and flight of steps; follow the path, over the stile, to Harley Bank. After 200yds on your right, at the edge of the field is a small exposure of Wenlock Limestone, with plenty of crinoid ossicles. Continue along the path, passing Stokes Barn (a recently renovated cottage and barn) on your left and enter the wood ahead by a small wooden gate. The quarry (long disused and overgrown, but recently cleared) is in front of you and to your left. Do not attack the quarry faces, as this would be both dangerous and unproductive, but examine the scree and the rock on the floor of the quarry. Persevere and you will discover plenty of typical Wenlock Limestone fossils - brachiopods, corals, crinoids, polyzoa, etc. Return to the car park.

J5. **Coates Quarry** (GR 605994)
Leave the car park, turn right and continue along the B4371 towards Church Stretton. After ½ mile you will see a large disused quarry on your right; this is known as Coates Quarry and you should be able to obtain permission to enter by asking at the works about 1 mile further along the road on the right. The quarry is, of course in the Wenlock Limestone and exposes both massive and reef limestones with numerous fossils.

J6. **Viewpoint** (GR 574968)
Leaving Coates Quarry, continue towards Church Stretton; the road runs along the top of Wenlock Edge. After about 3 miles there is a small layby on the right; park here. You are standing on Wenlock Reef Limestone. There are excellent views to the west over Ape Dale to the Stretton Hills.

J7. **Ippikin's Rock** (GR 568965)
About ¼ mile further on there is a National Trust sign; you can pull in here. (The Wenlock Edge Inn (GR 570963) is 250 yds south). Leave the parking area by the

gate and stile and go southwest across the field to the kissing gate. The view from the top of the Edge here (GR 569964) is justly renowned. Follow the stepped path for 25 yds, then bear sharp left up a small track; you will find yourself below the cliff of Ippikin's Rock. It is composed of reef limestones. Closer examination reveals thin layers of shale, which indicate periods when the coral reefs briefly ceased growing. Do not hammer here. Return to the stepped path and cross to the exposure immediately opposite. Here you will see bedded limestones, observably quite different from the reefs.

The National Trust path runs along the foot of Wenlock Edge for some 4½ miles to the car park near Much Wenlock. The first mile or so forms a Geological Teaching Trail, a guidebook to which is published by the Nature Conservancy Council.

J8. **Upper Millichope** (GR 521895)
Silurian mudstones, brachiopods, cephalopods, gastropods, graptolites and trilobites.

Tails of the trilobite *Dalmanites*

A world famous site for trilobites (*Dalmanites* and, less common, *Calymene*); brachiopods and cephalopods are also readily found and there are occasional gastropods and graptolites (*Monograptus sp.*).

Continue along the B4371 towards Church Stretton and after about 4 miles turn left to Rushbury. Drive through Rushbury, up and over the tree covered Wenlock Edge, fork right at the top and just over a mile further on is the locality, by the stream on your left near Upper Millichope Farm. Entry is via an iron gate and then right onto a small hillock of land, with trees. The stream flows between the land and the road and you should examine the exposures on the stream side of the hillock, but do not disturb the bank of the stream nearest the road.

J9. **View Edge Quarry** (near Craven Arms) (GR 426806).
See Itinerary D: South Shropshire Itinerary, Locality D4.

Refreshments:
The Wenlock Edge Inn. Superb site near Ippikin's Rock. Large car park and garden. Wide range of beers, snacks and meals. Tel. 01746 785678.

PHOTOGRAPHY

*Postcards; Greetings Cards;
Portraits - Family, Home, Pets;
School and Play Groups*

**Phillips Tutorials
Frogs Gutter, Minsterley, Shropshire. SY5 0NL
Telephone: 01588 650335**

GLOSSARY

What it all means and what to look for.
Where localities are given for finding the particular specimen, the letter refers to the itinerary and the number to the locality.

Fossils

A fossil is is an indication of past life found in sedimentary rocks. Human skeletons and artifacts are too recent to be described as fossils and are the concern of archaeologists and not geologists. The chances of finding the remains of an organism which was alive tens, or even hundreds, of millions of years ago, must be remote indeed and it is surprising that so many fossils are in fact found. Early life must have been abundant, especially in the sea.

To stand a chance of being fossilised an organism must undergo rapid burial after death, in order to avoid destruction by wave action, predators, etc. Fossils may be preserved in many ways: eg casts and moulds of hard parts; as carbon films (plants and some graptolites); replacement by mineralising solutions (petrifaction ie. "turning to stone"); as traces (trails, burrows, footprints); etc.

At the beginning of the Cambrian period (570 million years ago) fossils first appeared in great numbers and in a variety of types. To be fossilised successfully an animal must have "hard parts", ie. shells, bones, teeth, etc, and it would seem that it was around this time in the history of life on earth that animals with hard parts first appeared. One explanation for this event is that at this particular time the oxygen content of the atmosphere is believed to have increased greatly to a level similar to that of today, hence allowing sea creatures to form shells. However, vague traces of life, mainly algal forms, have been discovered much further back in time in the Pre-Cambrian rocks. Life appeared first in the sea (a relatively constant and amenable environment) and it was not until around 350 million years ago that life forms left the seas
and began to colonise the land.

Bivalves (also called Lamellibranchs)

Shellfish, like cockles and mussels. They have two irregularly shaped shells usually mirror images of one another. They are free movers, crawling along the sea bed or burrowing into it. There aren't many of them in early rocks - in fact, they are most abundant in the present.
Locality: G2.

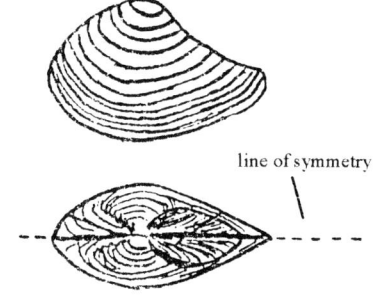

Brachiopods

Like bivalves these are also shellfish with two shells; but the shells are symmetrical in shape and often of very different sizes. Brachiopods were mostly anchored to the sea floor. They trapped food as it was washed along on currents. Brachiopds were a dominant life form in the early rocks and Shropshire alone has hundreds of species. These animals exist today, but are restricted to a small number of varieties living in tropical waters. 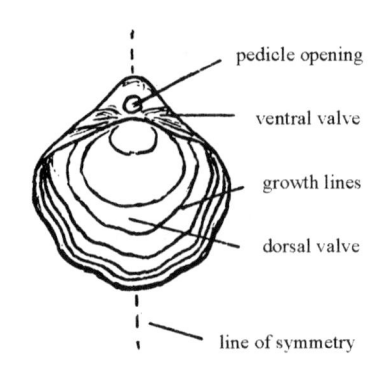
Localities: B2, B3, D2 - 4, E2 - 5, F4, G2 - 4, H5, J1 - 5, J8 and J9.

labels on diagram: pedicle opening, ventral valve, growth lines, dorsal valve, line of symmetry

Cephalopods

These were squid-like creatures that first appeared in the late Cambrian period. They are related to the better known ammonites and belemnites. The fossils are a tapered cigar shape, sometimes quite smooth and sometimes ribbed, like a bolt.
Localities: G2, G4, G6 and J8.

Corals

These animals are found as early as the Ordovician and survive today. They secrete a skeleton of lime, which is the part which is found fossilised, and this presents a circular or polygonal cross-section, with lines (septa) radiating out from a central hole (column). They may be solitary, usually horn-shaped, and quite large; or compound, small individuals living in colonies - like those of modern coral reefs. To discover a fossil reef in the limestone of Wenlock Edge is to enter a teeming sea of fossil life.
Localities: B3, D2, H5 and J1 - 5.

Solitary Coral 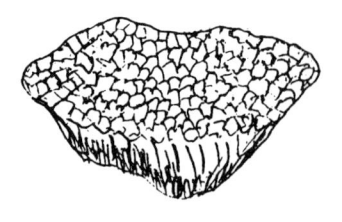   Compound Coral

Crinoids

Although commonly referred to as a "Sea Lily", this creature is not a plant but an animal, closely related to the echinoid (sea urchin). The calyx (head) is rarely found; the usual fossilised remains being short sections of the stem. The individual "washers" of the stem are called ossicles and in cross-section can be either circular or in the shape of a five-pointed star. To the upper surface of the calyx are attached food gathering arms and the base of the stem is fastened to the seafloor.
Crinoids range from the Ordovician to Recent, but few are found after the Cretaceous.
Localities: B3, D2 - D4, H5, J1 - 5 and J9.

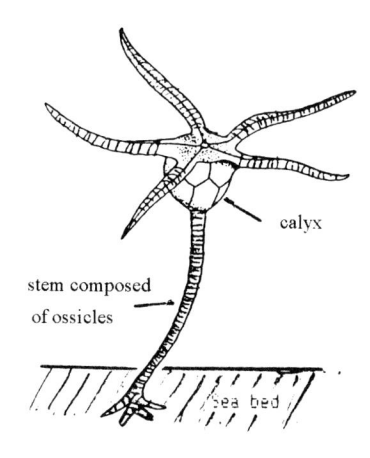

Fish

The first fish appeared at the end of the Silurian/beginning of the Devonian and were identified at Ludlow. The famous "Bone Bed", at times very thin, is usually a brownish sandstone, in which can be seen tiny black remains of the fish - scales and spines. Very important, but not a coffee table display item.
Locality: G1.

Gastropods

Creatures with a single, coiled, conical-shaped shell. These animals move on a large muscular "foot", appearing as their name suggests to "walk on their stomachs", like snails or whelks. Although primitive forms first appeared in the early Cambrian, gastropods only became plentiful, in both number and variety, in Tertiary and Recent times.
Localities: F4, G2, H5, J1 - 5 and J8.

Graptolites

This is a curious ancient life form that existed only from the very late Cambrian to the Devonian. They reached their peak in the Silurian. They were tiny creatures, perhaps somewhat like corals, living in colonies which floated in the ocean. The individual animals inhabited small cups (thecae), seen as a jagged saw-tooth edge, attached to a common stem (stipe). In fossil form they are

white, silvery or, in Shropshire, most frequently black traces on the rock; some-times confused with soil stains. The development of the creature was somewhat unusual, the animal appearing to have become "simpler" as it evolved; beginning as a multi-stiped form and ending as a single stipe with thecae on one side only (*Monograptus*).
Localities: G5, G6 and J8.

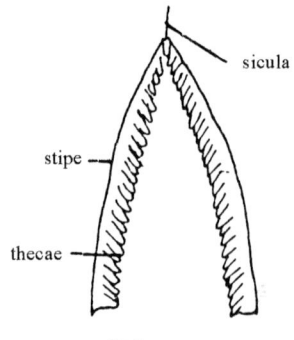

Didymograptus

Polyzoa (also called Bryozoa)
These animals are somewhat sinilar to corals, each individual inhabiting a minute chamber in a small stick-like or lace-like colony. They are known from the Ordovician to the present time.   Localities: F4, H5 and J1-5.

Trilobites
One of the most attractive fossils to the collector. They were like large wood-lice in appearance, living in shallow waters, prob-ably spending their lives crawling in the mud of the sea bed scavenging for food. The part of the animal usually preserved is the back or dorsal surface, which was covered by a chitinous exoskeleton and divided lengthways into three lobes, hence the name trilobite. There is also a transverse division into the headshield or cephalon, the central segmented body or thorax and the tailpiece or pygidium. The eyes, where present, were on top of their heads. Trilobites moulted periodically, casting off their exoskeletons and growing new ones; consequently isolated heads and tails are frequently found, whilst whole specimens of many species are comparatively scarce. They appear in the fossil record at the beginning of the Cambrian period and became extinct in the Carboniferous.
Localities: D3, E1, G4, G6 and J8.

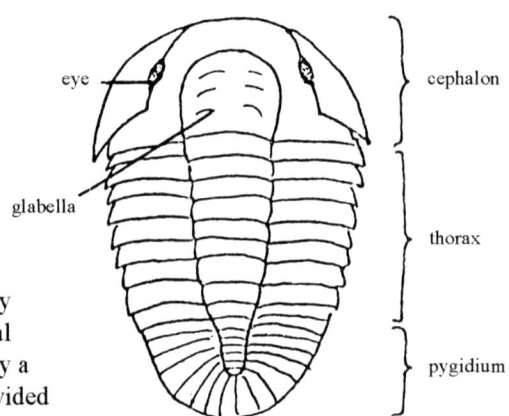

Glossary continued.

MINERALS and associated terms

Barite (Barytes)
Barium Sulphate. $BaSO_4$; white/colourless when pure, but often pinkish when traces of iron are present. Frequently occurs in tabular (flat plate-like) masses; it is relatively soft, like calcite, but is much denser (ie. it feels "heavy") and it does not fizz with acids. The pure white mineral was used extensively as a filler for paint and paper and even face powder. It was mined in Shropshire until the late 1940s and can easily be found on old mine waste dumps. Its main use today is as a constituent of the lubricating mud for drilling rigs.
Localities: C4 and C5.

Calcite
Calcium carbonate, $CaCO_3$; this mineral is the major constituent of limestones. When pure it is white or colourless and has a rhombohedral cleavage, ie. when a lump is broken it will form small rhombohedral shaped pieces. It is a relatively soft mineral (it can be scratched by an iron nail) and effervesces (ie. fizzes giving off carbon dioxide gas) when treated with acid - try vinegar: both these properties serving to distinguish it from quartz. Calcite is another common gangue mineral in metal ore veins and is therefore plentiful on many mine dumps.
Localities: C4 and C5.

Chalcopyrite (Copper Pyrite)
A copper/iron sulphide, $CuFeS_2$ and the most common primary ore of copper. It is similar in appearance to iron pyrite but is softer and has a deeper colour, frequently with a colourful tarnished appearance. Occurs in small amounts in some of Shropshire's mineral veins.

Cleavage
Because of their particular molecular structure, many minerals tend to split naturally in certain directions; eg calcite breaks into small rhombohedral-shaped pieces.

Galena
Lead sulphide. PbS; the most common ore of lead; very "heavy", silvery-grey

in colour with a metallic lustre on fresh surfaces but tarnishes rapidly. Galena forms cubic crystals, although such samples are not often found. More frequently a lump of galena can be seen to possess cubic cleavage, ie. breaks into pieces showing a cubic structure. The mining area of south-west Shropshire (Snailbeach, The Bog, Whitegrit, etc.) is known primarily for its production of lead, which has been mined intermittently since Roman times, reached its peak about the middle of the last century and ceased around 1900.
Locality: C4.

Iron Pyrite (Pyrites)
Iron Sulphide, FeS_2; its pale brass-yellow colour could possibly lead to this mineral being mistaken for gold, hence its alternative name "Fools Gold". A common sulphide ore, it is frequently present, in small amounts, in the lead/zinc veins of south Shropshire.

Malachite
A basic copper carbonate, $CuCO_3.Cu(OH)_2$; a secondary copper mineral and bright green in colour. (Azurite is chemically very similar, $2CuCO_3.Cu(OH)_2$, but bright blue in colour). Malachite and azurite occur as the result of the breakdown of Chalcocite (copper sulphide), another primary ore of copper. In Shropshire malachite, along with some azurite, is occasionally found in association with barite.

Mineral
A substance occurring naturally in the earth's crust; it is structurally homogeneous (the same throughout) and is composed of a small number of elements chemically combined together in a definite composition. Minerals may be identified by their distinctive properties, eg. crystalline shape, lustre, density, hardness, cleavage (direction of splitting), fracture (breakage in a direction other than that of the cleavage), streak (colour of the mineral in powdered form), etc.

Ore
Material worked for the purpose of extracting a metal and usually consisting of three parts: i) the Ore Mineral, from which the metal is finally extracted; ii) the Gangue Minerals, the "waste" minerals which accompany the ore mineral, but are not sources of metals, examples being barite, calcite and quartz; and iii) the Country Rock, the local rock enclosing the mineral vein.

Quartz

Silicon dioxide (silica), SiO$_2$; a very common mineral and a major constituent of many rocks. When pure it is colourless and transparent and can form crystals in the shape of beautiful hexagonal pyramids. More often however it is found in its white form ("milky quartz"); it is very hard and acid has no effect upon it. Quartz commonly occurs in veins, frequently with other minerals and is very common on mine waste tips.
Localities: C4 and C5.

Sphalerite

Also known as Zinc Blende or "Black Jack". This is zinc sulphide, ZnS, the primary ore of zinc; brown black in colour with a resinous lustre or sheen which is very distinctive. Usually occurs in association with galena. In the early days of lead mining, sphalerite was a waste material and it is therefore relatively easy to find samples on waste tips, whereas galena is far less plentiful.
Locality: C4.

Snailbeach Lead Mines

Reproduced by kind permission of Sylvia Hopkins

Glossary continued.

ROCKS and associated terms

A rock is a mixture of minerals forming part of the Earth's crust. Rocks can be divided into three groups - igneous, sedimentary and metamorphic - see below.

Andesite
A fine-grained, volcanic, igneous rock of intermediate composition; usually formed as a lava flow.
Localities: F3 and F4.

Bedding, Bedding Plane
A bedding plane is a surface parallel to the original surface of deposition of the rock, which in almost all cases will have been horizontal. Subsequent earth movements will have tilted these bedding planes. Shales tend to split along their bedding planes.

Conglomerate
A shallow water, or beach deposit. Pebbles and grains from near the shore became compacted into rock. Good examples are:
a) Stiperstones: bands of conglomerate occur, indicating that the sea at that time was invading and receding.
b) Silurian Basal Conglomerate: found south (near Hillend) and west (near Norbury) of the Long Mynd. This impressive rock is composed of pebbles worn away from the Long Mynd by the Silurian sea 400 million years ago. Its purple green and red pebbles represent the ancient shoreline, when the Long Mynd was the edge of the land.
Localities: B2, D1, D3 and H3.

Dip
The angle at which layers of sedimentary rock slope. Though rocks were laid down as sands, etc. in the ocean, later earth movements tilted them - or even folded them.

Dolerite
A medium-grained igneous rock formed just below the land surface. Dark green to black, it forms prominent landforms like the Breiddens and Clee Hill, were it is

quarried for road stone.
Localities: A1, B4 and H2.

Fault
A fracture in the earth's surface, produced by earth movements such as earthquakes and resulting in displacement of bedding planes and hence observable relative movement between the sides of the fault.

Flag (or Flagstone)
A well bedded, sedimentary rock, which breaks easily into slices like paving slabs.
Localities: A1, E2 and F4.

Flow Banding
Rhyolite frequently exhibits a colour banding, showing the flow of the lava prior to solidification.

Grit
A coarse sandstone, ie. made up of large sand grains.
Localities: B2, E5, F2 and F4.

Igneous Rocks
Formed by the cooling of molten material (magma) and are therefore crystalline. They may be extrusive, ie. formed on the Earth's surface, eg. rhyolite lava flows; or intrusive, ie. formed within the crust, eg. granite and gabbro. Igneous rocks formed at depth may be exposed by erosion.

Joint
A fracture in a rock, between the sides of which there is no relative movement. (Compare with a fault, where there is relative movement.) Joints may be caused by shrinkage - by cooling in the case of an igneous rock, or by the drying out of sedimentary rocks - or by tension in the rock due to folding.

Limestone
Rocks like chalk and Carboniferous limestone are formed from the shells and skeletal material of sea life. These are either compacted and/or dissolved and re-precipitated to form a sedimentary rock, the main mineral constituent of which is calcium carbonate. When a dilute acid is dropped onto a limestone effervescence (fizzing) will occur as carbon dioxide gas is evolved. The rock at Norbury is

somewhat unusual: while being a sandstone it has so many fossils in it that it reacts in many ways like a limestone.

Carboniferous or Mountain Limestone, such as at Llanymynech, is a common rock type and being tough and resistant is widely used for roadstone.
Localities: B3, D2, E3, F2, G3, H5, J1-7 and J9.

Metamorphic Rocks
Rocks, originally either igneous or sedimentary, which have been changed by the action of heat and/or pressure. Common examples are quartzite, slate and marble.

Mudstone
A compacted clay, which may disintegrate when wet, but does not form a plastic mass as clay does. It has no visible bedding and breaks into irregular lumps.
Localities: G2, G4-6 and J8.

Quartzite
A coarse sandstone which has been heated and compressed during earth movements, thus becoming very tough. It forms (the ridge of) the Stiperstones.
Localities: D1 and H3.

Rhyolite
A fine-grained, volcanic, igneous rock of acidic composition, ie. containing a high proportion of quartz. Rhyolite lava is very viscous and frequently shows flow banding.
Localities: F4 and H2.

Sandstone
A rock made up of grains of sand, ie. silica (quartz). Usually formed by deposition in water, but occasionally sandstones are formed on land, eg. the desert sandstone of Nesscliffe. Its colour varies according to the chemical impurities it contains; a white sandstone being almost pure quartz.
Localities: A2, B1, B2, D2, D3, E4, E5 and F2.

Sedimentary Rocks
Are formed from particles of other rock which have been worn away by water, wind and weather and deposited (laid down) in layers. Such rocks include conglomerates, grits, sandstones, mudstones, shales and limestones. Most sedimentary rocks are laid down in the sea. Those near the shore are made up of

large pebbles (conglomerates); further out in the ocean sand grains form grits and sandstones. The finest particles are found in the deepest oceans and form shales and clays. Thus by examining sedimentary rocks we can re-create ancient oceans and shorelines and map the geography of the past.
NB. Sedimentary rocks may contain fossils; igneous rocks do not. (Metamorphic rocks, such as quartzite, may have done, but no longer).

Shale
A mudstone (see above), which is clearly bedded.
Localities: E1, E5, F1, G2 and G4 - 6.

Tuff or Volcanic Ash
Small rock particles ejected from a volcano are referred to as ash; however when the ash has consolidated and formed a compact rock it is usually referred to as a tuff.
Localities: F1 and H2.

Unconformity
Rocks are formed (laid down) in sequence, the oldest underneath, the youngest on top. When a break, or gap, appears in the sequence, there is said to be an unconformity. Rocks can be missing for one of two reasons; either they were never laid down, eg. the area was above sea level for this period of time and no deposition took place; or they were formed, but have been removed by erosion, prior to the next layer being deposited.
Localities: E1, E5 and F5.

Vesicular
Volcanic lavas are described as vesicular if they contain small circular or oval shaped holes, produced by bubbles of gas trapped by the solidification of the rock. Where these holes have been filled with a secondary mineral the rock is said to be amygdaloidal.
Localities: F3 and F4

Notes